GALLIPOLI REVISITED

GALLIPOLI REVISITED

IN THE FOOTSTEPS OF CHARLES BEAN AND THE AUSTRALIAN HISTORICAL MISSION

JANDA GOODING

hardie grant books
MELBOURNE · LONDON

Published in Australia in 2009 by
Hardie Grant Books
85 High Street
Prahran, Victoria 3181, Australia
www.hardiegrant.com.au

All rights reserved. No part of this publication may be reproduced, stored in a retrieval system or transmitted in any form by any means, electronic, mechanical, photocopying, recording or otherwise, without the prior written permission of the publishers and copyright holders.

The moral right of the author has been asserted.

Copyright © Australian War Memorial 2009
Copyright photography © Australian War Memorial 2009
Cataloguing-in-Publication data is available from the National Library of Australia.
Gallipoli Revisited: In the footsteps of Charles Bean and the Australian Historical Mission
ISBN 978 1 74066 7654
Designed by Pfisterer + Freeman
Printed and bound in China by C & C Offset Printing Co.
10 9 8 7 6 5 4 3 2 1

Title page:
Hubert WILKINS (1888–1958)
Right of Leane's Trench showing the cornfield on top of the hill and Lone Pine in the distant centre, c. 26 February 1919
Glass whole-plate negative
AWM G01943 (detail)

Half title page:
Australian felt (slouch) hat
RELAWM00388

A slouch hat used by a 4th Australian Infantry Brigade soldier and found at Hill 60 by the Australian Historical Mission in 1919.

Wire entanglements
RELAWM00301

Found at Quinn's Post by the Australian Historical Mission in 1919.

CONTENTS

vi	DIRECTOR'S FOREWORD
vii	ACKNOWLEDGEMENTS
viii	NOTES ON TEXT AND IMAGES
ix	INTRODUCTION
1	THE RIDDLES OF GALLIPOLI
21	THE MISSION
43	TRAVELLING EAST
65	LAMBERT SETS TO WORK
93	WILKINS AND BEAN TRACE THE EVIDENCE
127	BATTLEFIELD LANDSCAPES: THE WORK OF JAMES AND SWANSTON
165	THE JOURNEY HOME
189	ART, PHOTOGRAPHY AND HISTORY
219	EPILOGUE
223	NOTES
230	SELECT BIBLIOGRAPHY
231	GLOSSARY OF ACRONYMS
232	CHRONOLOGY OF THE AUSTRALIAN HISTORICAL MISSION
234	INDEX

DIRECTOR'S FOREWORD

If anything has had a fundamental influence on the public's perception of the Gallipoli campaign, it is the 1919 Historical Mission to Gallipoli, led by Charles Bean, Australia's official war correspondent. The material that was collected and the work that was produced during the mission have come to form an important part of the Australian War Memorial's collection related to Gallipoli. Principal among these are Bean's notebooks, field sketches and measurements that map out the trench systems and the movements of Australian soldiers during the campaign of 1915.

An extraordinary group of men accompanied Bean on the Gallipoli mission, including the explorer, adventurer and war photographer Hubert Wilkins and the acclaimed Australian painter George Lambert. Wilkins's incisive and well-composed photographs show not only scenes of the old battlefield but also fragments of clothing, shell casings and human bones, which make a powerful statement of the tragedy and human loss. Lambert's small, brilliant panel paintings that were produced on the spot, and his subsequent large canvases *ANZAC, the landing, 1915* and *The charge of the 3rd Light Horse Brigade at the Nek, 7 August 1915*, have for many years helped to shape Australian perceptions of the landscape and the events associated with the failed campaign.

This book gives an overview of the Australian Historical Mission's journey across Europe to Gallipoli in the months immediately following the Armistice, and then its time on the peninsula. There they walked the ground and made the visual records that have become such an important part of the national collection. Through this story, the relationship between art, photography and history, and the development of the Memorial's collection in the early twentieth century, is clearly revealed.

This book is being published at a time when there is an increasing interest in the Gallipoli story and how it has influenced Australian identity. This is reflected in the increasing attendances at Dawn Services in Australia and overseas. I am sure *Gallipoli Revisited* will be of great interest to those wishing to know more about the Gallipoli campaign as its centenary approaches. I warmly commend Dr Janda Gooding for her splendid retelling of this fascinating story.

Steve Gower AO, AO (Mil)
Director
Australian War Memorial

ACKNOWLEDGEMENTS

This book has benefitted from the support and contributions of a number of people. Following the opening of the exhibition *George Lambert: Gallipoli and Palestine landscapes* in March 2007, the Director of the Australian War Memorial, Steve Gower, suggested that the topic could be expanded into a major publication detailing the 1919 Australian Historical Mission to Gallipoli. This idea was enthusiastically supported by Nola Anderson, Assistant Director, Branch Head National Collection, and Lola Wilkins, Head of Art. The continuing support of senior management at the Memorial has been crucial to the book's successful completion, and the author particularly wishes to thank Lola Wilkins for her encouragement over the last two years.

The Australian War Memorial facilitated much of the detailed research through a Major Research Program Staff Grant in 2007. Essential field work associated with this project was made possible by the Gordon Darling Foundation through a generous Global Travel Grant, which enabled the author to visit Gallipoli. Keith Mitchell prepared the maps and the Memorial would like to thank Mr E.B. Le Couteur and Mrs A.M. Carroll for granting permission to reproduce pages from Dr Bean's notebooks and diaries.

All publications benefit from careful and committed readers, and this one has been fortunate in the range of people who have contributed greatly to the final text. Thanks are owed to Steve Gower, Nola Anderson, Lola Wilkins, Ashley Ekins, Ian Affleck, Shaune Lakin, Diana Warnes, Bridie Macgillicuddy, Cherie Prosser and Kelda McManus. The Memorial's Senior Editor, Robert Nichols, fine-tuned the manuscript and made suggestions that have greatly improved the book.

The project was managed in the first instance by Linda Byrne and Stephanie Boyle, and later by Majella Edwards, with images prepared by the Memorial's Multimedia section, led by Hans Reppin. A special acknowledgement is due to Bob McKendry, who has skilfully worked from the original glass plate negatives to retrieve the clarity and beauty of the images produced by the photographers in 1919. Nick Fletcher, Joanne Smedley, Aaron Pegram and Andrew Jack have assisted with background research and captions for the images.

Lastly, it has been a pleasure for staff of the Memorial to work with the committed and professional staff at Hardie Grant Books. Sharon Mullins, the commissioning editor, has been a constant source of support, and Jane Winning has skilfully edited the text. Thanks also to designer, Hamish Freeman, who has produced an elegant and sympathetic design entirely compatible with the material.

NOTES ON TEXT AND IMAGES

Text

Throughout this book, short biographical endnotes have been provided for all of the principal military personnel when they are first mentioned. These notes have mainly been taken from military service records and indicate service in the First World War only. Rank and any decorations are given as they were at time of discharge. The use of such biographical notes follows a tradition established by Charles Bean in the official history series.

Place names are given as they appear in the official history of the relevant conflict. In some instances, when names have changed significantly since the time of the events being described, the modern form of the name is given in parentheses; for example, Constantinople (now Istanbul).

Titles for works of art are those given by the artist. In cases where the artist has misspelt a word in the title, this is not corrected, merely noted with [*sic*] to retain the artist's original intention and historical authenticity. Captions for photographs produced during the Australian Historical Mission in 1919 are based on those that appear in the original 'G' series caption book.

Please note that the Australian War Memorial always writes ANZAC in capitals, even when it is not being used as an acronym.

Images

All images reproduced in this book come from the Memorial's collection and are identified by their accession number. Captions accompanying the images contain information about the maker or artist, title and date. The place where a work of art was made is also included in captions.

Many of the photographic images held by the Memorial are in the form of the original negatives, with glass-plate or nitrate negatives featuring most prominently within the selection made for this publication. For the most part, the Memorial has a negative-based collection, and this has served to build a flexible image archive available for a range of uses.

There were no official Australian photographers on Gallipoli in 1915. Censorship on Gallipoli was not as strictly enforced as on the Western Front, and many soldiers carried pocket cameras into the field. The collection of photographic images that relate to Gallipoli would be significantly poorer without such photographs contributed by soldiers and their families. About half of the Memorial's holdings of Gallipoli photographs have come from these sources.

Herbert Baldwin, the first official photographer for Australia, was appointed in November 1916 and worked in France with Charles Bean until June 1917. He was replaced by Frank Hurley and Hubert Wilkins in August 1917, and shortly afterwards official Australian photographers were also working in Palestine to document the efforts of the Australian Light Horse units.

Since its beginnings, the collection has also benefitted from the generosity of donors who have loaned photographic material to the Memorial to make copy negatives. Donations of print-based photographs have been an important way for the Memorial to build its collections, and today the Memorial also collects 'digital' photographic images. Over 255,000 images from the photographic collection of nearly a million have been digitised and are available for viewing on the Memorial's website.

In preparing the photographic images for reproduction here, the Memorial's Multimedia section has worked directly from the original negatives to produce a digital image suitable for printing. Digital corrections have been kept to a minimum, but evidence of slight damage to the negatives caused by dust, scratches or even breakage has been carefully removed where such corrections do not alter the intent of the original image. Contemporary digital processes have been used to extract as much information as possible from the original negative and to reveal the extraordinary beauty and clarity of these images taken ninety years ago.

INTRODUCTION

Lunch was not yet ready to be served, so the boys drifted into the small museum just across the hall from the dining room. Young Charles Bean gazed at the walls and glass showcases filled with tantalising objects from the old battlefield of Waterloo. He was just nine, perhaps ten, years old, his two brothers even younger, but what they saw fascinated them: 'We youngsters steeped ourselves in the contents of the museum room'.[1]

Over two winters in the early 1890s, their father brought the family to holiday in the small but famous Belgian village of Waterloo. The Hôtel du Musée had once been kept by Edward Cotton, who as a sergeant-major in the 7th Hussars cavalry regiment had fought in the battle. As the main battlefield guide over many years, Cotton had collected relics that 'made the old battlefield live again'. Many years later, Bean could still clearly recall what he had seen in that small room: the tall peaked caps (or shakos) of Napoleon's Old Guard; the distinctive red coat of a British infantryman; old muskets, bayonets, ramrods and pouches; 'pieces sawn from old trees with round lead bullets in them'; even 'a few skulls with holes from round shot or bullets and clear sabre-chips'.[2]

With his father and younger brother, Bean also walked the battlefield itself, carefully noting the contours of the land where Wellington and Napoleon had arrayed their troops. They examined the ground for bits of corroded metal or small round shot: 'We used to pick up imagined relics in the fields around the farms of Hougoument and La Haye Sainte – probably bits of farm harness – to make our own museums.'[3] In tramping over the ground and looking for clues to its human history, the young boy was also taking his first steps towards gaining an understanding of the geographical and human dimensions of war. The powerful combination of landscape, history and objects brought the battle to life.

This understanding became a model that he would draw on years later, when he eventually came to found one of the best known museums in Australia.

The Australian War Memorial, Canberra, is a great Australian institution. It is also unique, as it combines the national memorial to Australian servicemen and women who have died in wars with a comprehensive museum that chronicles the Australian experience of war from the late nineteenth century to the present day. The idea for this museum was conceived on the battlefields of Western Europe during the First World War. Bean, one of the principal founders, was determined that Australia would have a world-class museum that would draw upon the wartime experience of individuals while placing the nation's contribution to the war within a larger context.[4] He envisaged a building full of all sorts of objects, moving stories told by those who had taken part in the war, and great works of art that would convey to the Australian public the power, trauma and tragedy of that terrible conflict. At the heart of Bean's vision for the future museum was a fundamental belief in the capacity of museum collections to have a lasting impact on how people comprehend major historical events.

Bean's background and early education were eclectic and not immediately suggestive of his later roles as author, historian and museum founder. Born in Bathurst, New South Wales, in 1879, he was educated in Australia and Britain before going up to Oxford to study Classics.[5] His letters to his parents, however, revealed an intense interest in military matters and he often illustrated these letters with lively sketches, caricatures and even the occasional map.

In early 1900 news of the South African War tempted him to seek a commission as a military engineer, but at the same time he was also vacillating between a career in the

Grave cross
RELAWM00426.002

Made from part of a biscuit tin, the raised lettering on the cross reads: 'In memory of No. 1243 Cpl D McVay (Yank) D Coy 23 Batt A.I.F. Killed in Action 13/9/15.' The cross was found on Gallipoli in April 1919 by Captain G.S. Keesing, an AIF officer, then attached to the Imperial Graves Commission, who later donated it to the Australian War Memorial.

Civil Service or the law. In the end law won out, if only temporarily. He returned to Australia in 1904 and began to write for various Sydney newspapers, quickly establishing a reputation for simple and honest reporting. As well as reporting on naval manoeuvres, he travelled extensively in the outback, and wrote substantial accounts of the wool industry in which he outlined the characters of the workers and what he saw as their intrinsic qualities. In 1910 he seized the opportunity to travel to London as the representative of the *Sydney Morning Herald*; there he reported on general events but also paid special attention to issues of imperial defence and the creation of the Australian navy. By the time he returned to Australia three years later, he had established wide-ranging contacts in the political, publishing and newspaper worlds and had gained respect for his coverage of military matters.

In September 1914, a month after the outbreak of the First World War, Bean won a ballot run by the Australian Journalists Association to become Australia's official war correspondent. Within two months he was on his way to Egypt with the first contingent of the Australian Imperial Force (AIF). He was with them when they landed on Gallipoli on 25 April 1915 and reported on events there until the evacuation in December. It was while he was on Gallipoli that he was reminded of the Hôtel du Musée near Waterloo and began to envisage a museum in Australia that might preserve the many artefacts and documents that would help to explain the Australian experience of war.[6]

During the next three years he moved back and forth along the Western Front with the Australian troops, all the time sending press despatches back to Australia, keeping detailed diary entries of military actions and making records of countless interviews. He continued to develop his idea for a museum and helped to establish a special unit called the Australian War Records Section (AWRS), which undertook a massive campaign to collect battlefield relics, documents and official records related to Australians at war. Also, under his guidance, an extensive commissioning scheme was set up to harness the skills of artists and photographers to create different representations of the war. By the end of the war

Brass bugle
RELAWM00415

Battered and shrapnel-damaged, this Australian brass bugle was found at Pine Ridge, Gallipoli, by Cyril Hughes of the Graves Registration Unit. He donated it to the Memorial in 1921.

Bean's vision of the museum had expanded, but his belief in the power of storytelling through images and original historical objects remained central to his idea for the new institution.

When an opportunity to revisit Gallipoli arose in early 1919, Bean quickly put together an extraordinary group of men to accompany him on what he called the Australian Historical Mission. The purpose of the Historical Mission was to develop a more complete understanding of Australian involvement in the military campaign against Turkish forces in 1915, and to collect and create material for the proposed national war museum. It was always important that the museum contain original artefacts, paintings and photographs as well as military records and to this end Bean selected a talented Australian artist, George Lambert, and a highly experienced photographer, Hubert Wilkins, to accompany him.[7] Working alongside the Historical Mission, two photographers from the Australian War Records Section in Cairo, William James and William Swanston, also completed a thorough photographic survey of the Gallipoli area.[8]

The task that Bean and the men of the Historical Mission undertook was immense, and their research, along with the relics, photographs, paintings and maps they compiled, has had a lasting impact on the way that Australians visualise and remember the 1915 ANZAC campaign. Notwithstanding its significant contribution to our understanding of the Australian experience on Gallipoli, the work of the Historical Mission has remained generally unacknowledged.[9] This book outlines the development of the Historical Mission and the work of its principal participants. It relates the reasons for the formation of the expedition and its aims, before tracing the group's progress across Europe to Turkey and then down to Egypt where it disbanded. However, the account would remain incomplete if all it provided was a chronology of a journey. Interwoven throughout the chapters is a discussion of the role of art, photography and history in the early development of the Gallipoli collections and subsequent displays of the Australian War Memorial.

Over nearly a century, the Australian War Memorial, with the enormous collection that Bean was so instrumental in forming, has remained committed to the power of historical objects to suggest a range of narratives to visitors. Detailed accounts of individual men and women are still used to illuminate the broader story of Australians and war, with each relic, each photograph and each work of art providing a clue to the history of Australia's military endeavours and to the responses of individual Australians during times of war.

Charles BEAN (1879–1968)
A view of the head of Mule Valley and Walker's Ridge, September 1915
Nitrate negative
AWM G00914

Bean's first response to the peninsula was astonishment: 'The sight of the hills as we got in closer and closer to see what they really were made one realise what our men had really done.'

CHAPTER ONE

THE RIDDLES OF GALLIPOLI

IN 1948 A MODEST BOOK WRITTEN BY CHARLES BEAN, *Gallipoli Mission*, was published by the Australian War Memorial.[1] Bean was a recognised authority on the subject of Gallipoli, having been there in 1915 as Australia's war correspondent, and he subsequently wrote the first two volumes of the official history, dealing with that campaign. He believed his new book about Gallipoli would help answer the many 'riddles' that were left unanswered when Australian troops evacuated their positions on Gallipoli in December 1915. Although many details of how the campaign had been fought seemed to be lost, the story would be revealed by deciphering 'marks left for several years on the ground' and by the examination of objects collected from the battlefield.[2] In this way *Gallipoli Mission* would pass on to fellow Australians 'experiences which seemed too interesting to be stored in a few fading memories and semi-official records.'[3]

These 'experiences' dated from almost thirty years earlier, when in 1919 Bean led the Australian Historical Mission to Gallipoli. This extraordinary group included a historian, an artist, a photographer, map-makers and relics collectors, and its main tasks were to gather evidence and make a visual record of the Australian experience on Gallipoli. Bean's book chronicled the participants' characters and the collective achievements of the group during its nine-week expedition. Informally written and driven by clear expression of both military history and personal narratives, *Gallipoli Mission* was an accessible alternative to the massive two-volume history of ANZAC that Bean had written in the early 1920s.

The publication of *Gallipoli Mission* prompted a Gallipoli veteran living in South Australia, Humphrey Kempe, to write to Bean.[4] It was a long letter and as well as congratulating Bean on the book, Kempe gave a powerful eyewitness account of incidents at the Nek, when on 7 August 1915 four lines of Australian troops had charged across a narrow piece of no man's land in a futile attempt to capture well-defended Turkish positions. On that day he had been attached to a machine-gun squadron and saw some of the tragic events from a distance. However, more than thirty years later, Kempe was still troubled by 'everlasting exasperating interesting but apparently insoluble questions' related to his time on Gallipoli.[5] Like many veterans grappling with reconciling things they had seen and done within the greater narrative of the ANZAC legend, Kempe saw in Bean's book an opportunity to revisit his experience and put some issues to rest. In concluding his letter, he wrote: '*Gallipoli Mission* is a great help, & will give one a much better chance to die in peace now the riddles have vanished.'[6]

When the ANZAC forces left Gallipoli at the end of 1915, many details of the campaign remained unclear. Indeed, some would prove to be unknowable. Record-keeping during the campaign had been inadequate and official documents were later described as 'scrappy' and insufficient to prepare

David BARKER (1895–1946)
Gallipoli (cover illustration for *The ANZAC Book*), 1915
Watercolour and ink on paper, 32.5 x 21 cm
AWM ART90422

a detailed military history of the events from April to December 1915.[7] A primary objective of the 1919 Historical Mission was to try to resolve many of the questions that remained about the ANZAC campaign. Who had gone farthest inland on the day of the landing? What had really happened when those four lines of Light Horsemen charged across the Nek? Where were the Turkish guns (nicknamed 'Beachy Bill') that had fired on ANZAC troops? And crucially, what had the Turks seen from their positions, and what could the allies have seen had they captured them? These and many other questions had plagued the soldiers during 1915 and still remained to be investigated in 1919 when, with the end of the war, it became possible to revisit the old ANZAC area.

The importance of the Historical Mission to Gallipoli lies not just in the accessible account Bean gave in *Gallipoli Mission*. Almost anyone researching Australian involvement in the Gallipoli campaign of 1915, and its subsequent place in the nation's history and popular imagination, must eventually

Maker unknown (printed in Britain)
A souvenir of the Great World War and the glorious part played by Australia and New Zealand. The landing of our gallant sons of empire on the Gallipoli peninsula, 1915
Printed cotton scarf, 47.3 x 65 cm
AWM REL/08693

find their way to the collections of the Australian War Memorial and, in particular, to the collections formed in 1919 by the Historical Mission. Comprising battlefield artefacts, photographs, works of art, diaries, personal correspondence and sketch maps, the material gathered by this group is a central component of the official National Collection related to Gallipoli. Brilliant little paintings by the artist George Lambert, and his large commissioned canvases of the ANZAC landing and the tragic events at the Nek, are permanently on view in the Memorial's Gallipoli galleries. The work of official photographer Hubert Wilkins, and that of the AWRS photographers William Swanston and William James, who worked alongside the Historical Mission, can be found throughout the galleries, in official histories and in other major publications on the campaign. And Bean's personal archive of notebooks, diaries, photographs and sketches constitutes a substantial part of the Memorial's core collection. Ninety years after the Historical Mission visited Gallipoli, its work continues to shape the answers to many of the riddles that still linger around this part of Australia's history.

ANZAC, THE AFTERMATH

Prior to the First World War the Australian public's understanding of other parts of the world was framed by strong personal and political ties to Britain and the empire. Conflicts such as the Boer War (1899–1902) and the Boxer Rebellion in China (1900–01) had cemented Australia's responsiveness to the call of empire. Involvement in the First World War, however, propelled the newly formed Australian nation into the international arena. Loyalty to the British Empire was tested again, but in the process new and independent associations were formed with places that had hitherto impinged little on the consciousness of Australians but that subsequently became inextricably woven into the national story.

Artist unknown
Australia's imperishable record, c. 1915
Colour lithograph on paper, 95.7 x 64.6 cm
AWM ARTV00850

From left to right: The Reverend Edwin Bean; Charles Bean, Australia's official war correspondent; Archie Whyte, editor for *The Age*; and Phillip Schuler, correspondent for *The Age*. This photograph was taken just before Bean and Schuler embarked with the first contingent of the AIF.

Photographer unknown
Standing on the dock at Port Melbourne,
c. 21 October 1914
Gelatin silver print
AWM A05379

Phillip SCHULER (1890–1917)
Charles Bean in the long communications sap that ran from ANZAC Beach, through No. 2 Post and up to the New Zealand headquarters, 26 July 1915
Nitrate negative
AWM PS1580

Astonished at just how quickly the landscape had been transformed, Bean commented that 'The post was taken up about three weeks after we came here. At the time there wasn't a trench outside of it – now the whole place is fortified and entrenched.'

When Australian troops landed on Gallipoli on 25 April 1915 Australian press coverage tended to highlight Australia's contribution within the overall imperial endeavour. On 1 May, as the first news from the Dardanelles was printed in Australian newspapers, King George V congratulated the country on the performance of its troops, who had 'proved themselves worthy sons of the Empire'.[8] But the reports emanating from Bean, Australia's official correspondent, focused less on empire and more on what he saw as the distinctive and fearless character of Australian troops as they went into battle commanded by 'gallant officers'.[9] His despatches over the course of the war were based on personal observation as he recounted the stories of individuals working in extreme conditions. Bean often leavened tragedy with humorous anecdotes, but he was always governed by a desire to report everything in a factual and objective way. He believed that Australians wanted accurate reporting of events and although at the time his writing style was criticised for being bland, Bean's philosophy as a war correspondent was to 'err on the side of care' rather than to sensationalise for the sake of journalistic appeal.[10] As an employee of the government, he saw himself as the 'official' correspondent for Australia and so not beholden to any particular newspaper. However, Bean took great care never to include anything that was critical of the government and generally made his reports 'unexceptional' so as to meet with the official censor's approval. In this respect, Bean was a willing participant in the machinery of war, and therefore careful not to contravene the boundaries established by official censors.

Charles BEAN (1879–1968)
Australian troops inside a captured Turkish trench at Lone Pine, 12 August 1915
Nitrate negative
AWM G01126

Bean went up to Lone Pine not long after its capture and made note of the construction of the trenches with rough pine logs. In 1919 some of these logs were selected by the Australian Historical Mission and brought back to Australia, where they are now displayed in the Gallipoli gallery at the Australian War Memorial, Canberra.

Bean rushed up to Quinn's following a Turkish attack and, while sitting with the troops, had a brush with death: 'Down the path came rolling an innocent black ball, like a cricket ball. It reached [a] dead Turk about 6 feet from me then exploded like a big Chinese cracker ... I was spattered over with bits of dead Turk – fortunately, not thickly.'

Charles BEAN (1879–1968)
Troops of the 4th Brigade in a communication sap behind Quinn's Post, 29 May 1915
Nitrate negative
AWM G01011

As a war correspondent, Bean was an acute observer and recorder of what he saw and heard. He gained the confidence of senior commanders but did not rely solely on them for information. He was also able to move around freely, visiting the front line to see what was happening or seeking out men to ask questions about the day's events. Fascinated by people and keen to hear about their experiences, Bean became increasingly well known and respected among officers and troops alike. His understanding of the campaign was therefore formed through this combination of personal experience and first-hand accounts from regular soldiers. As a result, when reporting on Gallipoli, and later on the Western Front, he was often close to the fighting and sometimes at considerable risk.

Like many others, Bean was conscious of the historical importance of Australia's involvement in the Gallipoli campaign. He was also aware that he was in a privileged position and that his personal records would be valuable in their own right. As he steamed towards Europe with the first contingent of the AIF in late 1914, he decided he would use his diary as the chief form of recording the war as he saw it.[11] In entries that were often written late at night, he recorded detailed information about the people and events he observed. These notes were essential for his press reports as well as for the history he was asked to write when the war ended.[12] Of the 296 diaries and notebooks that he compiled between 1914 and 1918, about 25 were devoted to the events on Gallipoli and many others contain interviews with Gallipoli veterans he later interviewed on the Western Front.

Phillip SCHULER (1890–1917)
Charles Bean at sea, 28 November 1914
Nitrate negative
AWM G01561

Word came that the AIF was to stop and train in Egypt rather than travel on to Europe. Bean was excited by the prospect of going ashore and 'got into camp rig at once'.

Designer unknown (Rotary Photo Company, Britain)
A Gallipoli souvenir. Bravo ANZACS!, c. 1915
Souvenir postcard, 14 x 9 cm
AWM RC00061

Meticulously following his formula of recording personal observations and verifying details using a number of sources, Bean's cautious approach and reliance on corroborated evidence may well have stemmed from his training as a lawyer and practice as a journalist.

Bean was on Gallipoli for almost the entire campaign, from the landing to the evacuation.[13] In his diary of 20 December 1915 he described the evacuation of the ANZACs as he saw it from the deck of HMS *Grafton*. He kept an all-night vigil, noting the time of every sound and movement on the shore, the bombs going off along the ridge lines, and the occasional firing across the lines by the last few troops. The night was full of anxiety, but Bean's account is tinged with melancholy as he describes the familiar posts and ridges gradually being deserted. At 3.20 am he noted that Quinn's Post and Pope's Hill had been abandoned, making 'an extraordinary end to a fine history. The Turks have got it at last – the place they never could take – by our quietly leaving it in the night.'[14] As he wrote his last few notes in the dark, he observed that 'perhaps the greatest success we have achieved there is quietly giving it to them without their knowing it.'[15]

The official estimate of Australian deaths during the Gallipoli campaign is 8,709.[16] As the nation's first major experience of modern warfare, and despite being a tragic failure, the events on Gallipoli rapidly assumed an important place in Australian political and cultural life. The troops were hailed as 'heroes' who had battled overwhelming odds to gain a foothold on enemy soil. On the first anniversary of the landing, Senator George Pearce, the Australian Minister for Defence, wrote of their achievements, invoking their pioneer settler ancestors who had torn themselves 'from homeland firesides to shape careers in this great island continent, and to overcome with indomitable pluck the awful hardships of a pioneering life.'[17] In this scenario, ANZAC troops were part of a natural progression of historical 'types' that began with the first settlers, then moved on to goldminers and bushmen. Through the fighting on Gallipoli Australians had earned their place on the world stage: 'For generations to come the story of the entry of the Australian troops to the European battle-field will ring in the ears of English-speaking nations.'[18]

Charles BEAN (1879–1968)
The landing, 25 April 1915
Nitrate negative
AWM G00894

Charles Bean took this photograph in the morning light on the day of the ANZAC landing. It shows boats carrying members of the 6th and 7th Battalions leaving the *Galeka* (right) and boats carrying the 2nd Brigade Headquarters leaving the *Novian* (centre).

In France in 1916, Australian Prime Minister Billy Hughes evoked Tennyson's famous poem 'The charge of the Light Brigade' (1870) when he addressed Australian troops assembled on the Western Front:

> On this day, called Anzac, one short year ago, the Australasian soldier leapt unheralded into the arena of war, and by a display of courage, dash, endurance, and unquenchable spirit proved himself worthy of kinship with those heroic men who throughout the history of our race have walked unafraid into the jaws of death, thinking it glorious to die for their country.[19]

Just two years later, Hughes cited Gallipoli as a key moment in Australian history, observing that the day of the landing at ANZAC Cove was the day the Australian nation was born: 'Before that we were New South Welshmen, Queenslanders, Victorians, – but on that day we became Australians.'[20]

Alongside the political rhetoric that was starting to shape the idea of Australia as an independent entity, Australian families mourned the dead and tried to reconcile their experience within the greater national narrative. The trauma of the war may have involved Australians in events larger than their immediate and parochial concerns, but the legacy of personal loss was real and located in homes across the country. Grief over loved ones killed on Gallipoli remained palpable in many communities, as people tried to find ways to deal with their sorrow.

A decision in 1916 to not repatriate bodies to Australia meant that bereaved families had to mourn in the absence of physical remains, sometimes without knowing the location of the grave or the exact circumstances of the death. Many people had a profound need for these details and saw visiting the battlefields and cemeteries as a way to reconcile their loved one's death within the bigger picture. However, the high cost of international travel limited the number of Australians who could go to Europe or Turkey, so some mourners found solace in objects collected from the front line, personal stories from other soldiers, or photographic images of battlefield landscapes and individual gravesites. Replete with a rich and traumatic history, these keepsakes not only helped people understand the detail of military events, they also stimulated discussion and facilitated the grieving process.

BEGINNINGS OF THE NATIONAL COLLECTION

From the first moment he waded ashore on Gallipoli, around 10 am on 25 April 1915, Bean took photographs and occasionally made small sketches of his surroundings. He was not alone in his determination to mark the moment. Many soldiers had landed with pocket cameras and the photographs they took were sent home and circulated widely among families and sometimes published in the Australian press. Others, such as Ellis Silas, made drawings of their surroundings and experiences in sketchbooks, and the landscape artists George Benson and Horace Moore-Jones

Photographer unknown
The crew of HMAS *Australia* listen to an address by the Australian Prime Minister, William Morris 'Billy' Hughes, Devon, 21 May 1916
Glass half-plate negative
AWM EN0080

Charles BEAN (1879–1968)
Headquarters staff of the 1st Division wade ashore at Gallipoli on the morning of 25 April 1915
Nitrate negative
AWM G00903

assisted commanders by making accurate studies of the topography.[21] When soldiers left, they took with them souvenirs of their experience and much of this material created on Gallipoli and collected from the battlefields eventually made its way to Australia.

Although he did not mention it in his Gallipoli diaries, Bean is believed to have discussed with Australian headquarters staff on Gallipoli in 1915 the possibility of creating a 'museum in Australia in which war relics would be preserved'.[22] At the start of the war Australia had no system for collecting relevant military records; the collection of all official records and the subsequent distribution of any objects collected from the battlefields was the responsibility of the British authorities. The establishment of the Canadian War Records Office in January 1916 meant there was increasing competition from all the allies to salvage material destined for museums. Knowing that Britain was about to form its own war museum, in November 1916 Bean raised the issue with Senator Pearce, noting the need to preserve objects for a future Australian war museum.[23]

While a formal decision on such a museum was not immediately forthcoming, in May 1917 the AWRS was established as part of the AIF. The officer-in-charge was Captain John Treloar and his brief was to undertake the systematic collection of material that would help document the war from an Australian perspective and to collect objects to form what later came to be called the National Collection of the Australian War Memorial.[24] Treloar was assisted by staff who would 'preserve and tenderly care for the sacred things which will some day constitute the greatest public possession Australia will have'.[25] Additionally, from mid-1917 onwards Australian officers were urged to compile more comprehensive unit diaries, and troops were encouraged to collect interesting items from the battlefields on behalf of the AWRS. The campaign was enormously successful and captured the imagination of many Australian soldiers, to the extent that Australian troops soon gained a reputation for scouring battlefields for relics. By the end of 1918 the AWRS had swelled to employ over 600 military and civilian staff.

Importantly, the conception of what would constitute the collection was not limited to documents or traditional war trophies. Australia's records would also include a pictorial record made up of photographs and works of art. These different forms of visual representation would give multiple perspectives and interpretations of people, places and events. In 1917 Bean described the nature of the collections being formed by the AWRS:

> There are three records, the written record – that is to say, official diaries, memoirs, maps, orders, messages sent and received in battle, and official or unofficial correspondence; the record of pictures – official photographs and cinema records, unofficial sketches, and the official artist's pictures painted or drawn on the spot; lastly, the records in material – trophies, the things our men and units have actually used in battle or the German has used against them, with the marks of battle upon them.[26]

To gather the photographic record of Australians at war, Bean argued that Australia needed its own photographic unit rather than relying on access to British official photographers.[27] In August 1917, Australians Hubert Wilkins and Frank Hurley, both experienced photographers, were appointed as official photographers for the AWRS.[28] Bean made it clear that the new photographic unit had two distinct functions. Hurley was to be responsible for taking photographs for press and publicity purposes, while Wilkins would provide a photographic inventory of 'actual conditions – the grim and sordid along with the heroics'.[29] Wilkins's images would be used as the evidence of Australian actions in events important to the nation's history and would document 'the scene of every tough struggle, the trench corner or blockhouse that proved a knotty point, the hills that overlooked us, and pictures taken from them showing what the German could see'.[30]

THE RIDDLES OF GALLIPOLI

15

Friends of those who had died at Gallipoli tended the various cemeteries with great attention in the final months of the campaign. As the withdrawal from the peninsula approached, Bean recognised that for the men 'the leaving behind of these graves hurt more deeply than any other implication of the Evacuation'.

Charles BEAN (1879–1968)
Queensland Point Cemetery, 1915
Nitrate negative
AWM G01292

Charles BEAN (1879–1968)
The silver lining – sunset over Imbros as seen from ANZAC (illustration for *The ANZAC Book*), Turkey, 1915
Watercolour and pencil on paper, 27.1 x 38.4 cm
AWM ART00044

In addition to the photographers, artists were employed to provide a different dimension of visual testimony of the war. In December 1916 Australian artist Will Dyson applied to the Australian High Commission in London and was granted approval to travel to France to work with the AIF.[31] Dyson's ensuing friendship with Bean, and the extraordinary work he produced, was a catalyst for the creation of an official war art scheme. Over the next three years the 16 Australian official war artists operated in two distinct programs to produce emotional and artistic interpretations of what they saw of Australians at war in Europe and the Middle East. One program managed by the AWRS selected serving soldiers who had had art training; they were to work with camouflage units and then to make sketches of the everyday experience of the soldiers around them. In a second scheme, artists were selected and managed through the office of the Australian High Commission, London, with support in the field from the AWRS. High-profile professional Australian artists were appointed and worked closely with specific AIF units to make sketches and small paintings of military actions with a view to later working these up into large-scale commissioned paintings. These paintings came to form the core of the nation's official war art collection. Bean and Treloar were involved in both schemes, so there was considerable overlap of personnel in the management of the official war art scheme.

The work conducted by the AWRS on the Western Front throughout 1917 and 1918 helped to form a comprehensive record of the contribution of Australians to the war. An American writer even said that the Australian record gathering project had 'developed scientific efficiency'.[32] Bean also spent his time working on the battlefields, recording and reporting on the Australian effort. But trying to fill in the many details that remained unclear about the Gallipoli campaign remained a high priority for him. From 1916 to the conclusion of the war, he continued to seek out and interview men who had served on Gallipoli as he pursued a more complete understanding and record of the actions of 1915. His diaries from this period are filled with his extensive notes and sketches from these conversations.

Australia's casualties on the Western Front were immense compared with those on Gallipoli, but the earlier campaign had come to occupy a central presence in Australia's wartime experience and there was a pressing need to understand exactly what had happened in 1915. Bean felt compelled to revisit the place as soon as possible. A second visit would help him prepare for his work on the official history of Australia's part in the war. He also wanted to walk the ground and to solve the many riddles that still existed. This would fulfil a promise that he and other veterans had made to 'visit the Turkish lines at Anzac & see for ourselves the trenches & country behind their lines'.[33]

In 1918 the Australian government approved the concept for a national war museum. The systematic collection of material from the Western Front had already been undertaken by the AWRS. From early 1918 this was matched by a campaign in Australia which asked the next of kin of servicemen and women to consider donating important historical items to a national collection. Eventually, these would be displayed in the new museum, thus creating a permanent record of the AIF's contribution to the war.[34] The comprehensive representation of the ANZAC campaign on Gallipoli in the future museum remained problematic, however, owing to the inadequate nature of the official records and paucity of material collected at the time. Bean had picked up a few artifacts while on Gallipoli; these were stored in Cairo for safe-keeping, but the hasty evacuation of ANZAC forces in December 1915 meant that much important material had to be abandoned. Unofficial photographs taken on Gallipoli were relatively plentiful because of the less stringent censorship controls operating there in 1915, and personal diaries and artworks made by soldiers would certainly assist in representing aspects of the campaign. The future museum, however, would have to rely on the generosity of donors to create a substantial national collection, and it was clear that unless immediate steps were taken, Gallipoli would be poorly covered in its displays.

Bean based his despatches on his own observations and experiences: 'In these deep, narrow alleys the front-line troops and supports lived as completely enclosed as in the lanes of a city, having their habitations along them in niches undercut in the wall, sometimes curtained by hanging blankets or waterproof sheets … The bivouacs on slopes behind the lines resembled the clustered booths at a great fair.'

Charles BEAN (1879–1968)
A general scene of Steele's Post at ANZAC, July 1915
Nitrate negative
AWM G01076

Photographer unknown (printed in Britain)
Memories of Gallipoli. On the way to the firing line. On April 25th 1915 Australasian history began, c. 1915
Souvenir postcard, 9 x 14 cm
AWM RC00058

This composite photograph is also known as 'A hop over' and 'A raid'. According to Hurley, it was impossible to capture the awesome scale of a modern battle 'without resorting to composite pictures', but this approach clashed with Bean's desire for accurate reporting.

Frank HURLEY (1885–1962)
Over the top, c. 1918
Glass half-plate copy negative
AWM E05988A

CHAPTER TWO

THE MISSION

ON 12 NOVEMBER 1918, THE DAY AFTER GERMANY SIGNED the Armistice that ended the war, Charles Bean was in London to seek formal approval to assemble the Australian Historical Mission and return to Gallipoli to complete a thorough survey of the site. He chose this name for the group in an attempt to give it official status and 'a little standing with those with whom we have to deal'.[1] As historian Bill Gammage has noted, the use of the quasi-religious term 'mission' may be a key to the way that Bean approached the project. It was something he felt destined to do. The return to Gallipoli gave him a chance to 'pay homage to the Anzac dead, and to use their faith and fire to inspire his fellow Australians to national greatness'.[2]

The structure and aims of the Historical Mission were also part of a tradition that had its antecedents in eighteenth- and nineteenth-century natural history and collecting expeditions, which had aimed to understand the natural world by forming a catalogue composed of scientific knowledge and images of the landscape. The creation of a visual record was an important function of these expeditions, and artists trained in scientific, military and topographic drafting were often included to make sketches of the people and places encountered.

The invention of photography in 1839 offered new opportunities. It was a technique that many believed could objectively capture and convey information with scientific precision. Unlike drawings or paintings, photographs did not have to be re-interpreted into other mediums for publication and the exact replication of a photographic image through the printing process meant that information was not corrupted when transmitted. But it was not until cameras became more reliable and portable that photographers were routinely included in scientific expeditions.

In 1918, when Bean proposed the idea of the Historical Mission to Gallipoli, he was continuing this tradition of a scientific collecting expedition. With himself as principal investigator and, accompanied by an artist and photographer, the Historical Mission would thoroughly research, record and collect any traces of war that remained on Gallipoli.

TRUTH, ART AND PHOTOGRAPHY

From the outset, the concept of 'truth' in the reporting and documentation of Australia's involvement in the war guided Bean. Truth and accuracy were at the heart of his own work as both correspondent and historian and these were also the guiding principles for the formation of the pictorial record. Bean was not alone in this search for the most appropriate ways to accurately represent the human experience of modern warfare. In Britain, tension between the government's desire to report events quickly and candidly, while at the same time protecting national interests, frequently erupted in debates about censorship, the use of propaganda, and the public's access to information. During the war, art and photography were harnessed by governments on both sides for propaganda purposes and both were subject to censorship at various times. In 1917 and 1918 war art and photography became a topic for discussion when exhibitions displayed in London generated considerable argument about the capacity of each medium to produce a 'true' account.

Photography possessed an immediacy ideally suited to the reporting of war. Improved technologies meant that an image could be taken in the field one day and appear in newspapers and magazines the next. It was considered a scientific medium and many people believed that the camera was just a mechanical apparatus that faithfully represented the subject in front of the lens. Most people didn't realise that it was still the result of one person's choice of subject combined with multiple individual decisions that informed the making of the image. However, the public's confidence in the reliability of photographic images was shaken when it was revealed that some war photographers re-staged events. In reporting the war, Australian official photographer Frank Hurley created composite photographs that challenged the notion that 'the camera couldn't lie'. These images were the result of combining elements from a number of negatives into a single image to achieve a dramatic synthesis of a scene that was not otherwise possible. In Hurley's opinion the awe-inspiring nature of war could never be conveyed in a single image, and so his composites were intended to present a dramatic and comprehensive overview of battle. The idea of the composite was not new, having been used during the American Civil War, but its use by Hurley tested Bean's desire for scrupulous honesty in all aspects of Australia's war records.[3]

On the other hand, the very foundation of art was that artists mediated their experience of the world using a range of technical and emotional approaches, conjuring up a different interpretation of reality that could be shockingly raw and intense. In early 1918 this was illustrated by an exhibition of war art by C.R.W. Nevinson, one of a number of British artists who had spent time at the front. His modernist paintings, when shown in London in 1918, provided visual testimony of the brutality and ugliness of modern warfare. Yet it was as much the modern style employed by the artist as the sordidness of the scenes that shocked the public and prompted a debate about truth and censorship. Bean would have been aware of these debates, as issues of truth and accuracy were central to his concept of the collection of war pictures and photographs.

Bean was not particularly impressed with the British war art scheme (of which Nevinson was a member) and he wrote rather condescendingly to his parents that the British had employed artists 'of the fashionable sort to paint their national pictures'.[4] Similarly, the Canadian scheme had produced paintings, at great expense, that were interesting but a mix of 'curious styles of contemporary art'.[5] Bean was proud of what had been achieved for Australia with a modest budget and using only Australian artists, believing that the pictures produced were a 'far more interesting set, & a suitable memorial' to the work of the Australian forces during the war. The public would find them to be engaging, accessible and accurate pictures that had been produced by 'enthusiastic men doing their best to help their country's record'.[6]

From its conception, the future war museum was envisaged as a way of commemorating the achievements of Australians at war. As such, it was important that it have multiple forms of representation – art, photographs, objects – to provide as comprehensive an account of the war as possible. In their unique ways, each form of record would give a different but 'truthful' interpretation of events to viewers. Bean expected that the artist and the photographer selected to accompany him on the Historical Mission would produce the official art and photographic record of Gallipoli to the exacting high standards he and others in the AWRS had already established for the collection.

To make this composite photograph Hurley combined the negative (E01201) of a photograph of Australians who had been wounded during the attack on Passchendaele village with a dramatic cloudscape (E05430D). For Bean it was a constant struggle to keep Hurley from retouching, staging and 'faking' images for greater effect.

Frank HURLEY (1885–1962)
Retaliation Farm Dressing Station, c. 1917
Copy negative
AWM P04060.005

Unknown Australian official photographer
Charles Bean (left) with Captain Henry Casimir Smart (right), Officer in Charge of Military Records, AIF Headquarters, at a sandbagged dugout at Montauban,
December 1916
Glass half-plate negative
AWM E00064

THE HISTORICAL MISSION MEN

Bean selected experienced men to work with him. His photographer, Hubert Wilkins, was the official war photographer with whom he had worked very closely on the Western Front. The artist, George Lambert, had already served as an official war artist with the Light Horse in Egypt, Palestine and the Sinai in early 1918. These were men he knew and trusted, and the photographs and works of art they produced during the expedition were to form the nucleus of an official collection related to Gallipoli. Most importantly, their work supported Bean's concept of truth in war records work.

Wilkins's background made him an ideal choice as the official photographer. He was born in 1888 in rural South Australia, in the harsh rain-shadow country of Mt Bryan East. He had an interest in engineering and mechanics, and had worked in the 'moving pictures' industry and then in aviation. He already had personal experience of war as a correspondent in the First Balkan War (1912–13), where he had used his skills as a cameraman and photographer to take what is now believed to be the first film footage of front-line action in a war.[7] Wilkins had also been part of Vilhjalmur Stefansson's Arctic expedition of 1913–16, and it was while in Alaska in 1916 that he heard of the war in Europe and made his way to Australia to enlist as a pilot in the Australian Flying Corps (AFC). He arrived in Britain in July 1917, but a sight test indicated he might be colourblind and therefore unsuitable as a pilot. At about the same time, because of his background as a war reporter and photographer he was recommended to Bean as a suitable candidate for the newly formed AWRS photographic unit.[8]

After his appointment as official photographer in August 1917, Wilkins worked closely with Bean to make a photographic record of the men, battles and landscapes of the Western Front. Apart from briefings he received from Bean, Wilkins's role was to roam across the battle lines and record Australians in action and at rest. Seeing men emerge dazed and exhausted from the front line, Wilkins came to the conclusion that '[h]uman beings seemed insignificant in the midst of all this. It didn't seem possible that men could go through it and live'.[9]

Despite the horror all around him, Wilkins kept his youthful belief that he would not die by misadventure. This belief had been tested a few years earlier in the Balkan War when he was arrested as a spy and reportedly faced a firing squad three mornings in a row.[10] Working on the Western Front, his fearlessness and intuition that something important was happening frequently took him up to the front line and on one occasion beyond it. In April 1918, while he and his camera assistant were following the fighting around Strazeele, Flanders (now France), they found themselves on top of a hill looking down onto a village full of what appeared to be French – but were in reality German – soldiers. They were spotted, and artillery shells started to land around them, but Wilkins insisted on taking a picture before picking up the camera and scampering about a kilometre back to safety.[11] He was well known for putting his own life at risk to obtain a photograph and his actions during the war earned him several commendations as well as a Military Cross and Bar.

While Wilkins's boldness in obtaining photographs had quickly earned Bean's respect, it was his integrity and diligence that Bean most admired. For Bean, Wilkins was the record photographer par excellence – he believed his photographs could never lie. As he said, 'they are a sacred record – standing for future generations, to see for ever the plain, simple truth'.[12] Wilkins was his preferred photographer, and it was his approach that he wanted for the photographic inventory of Gallipoli. From Wilkins's perspective, Bean was a man of honour and courage who provided an outstanding example of commitment and energy to those who worked around him. Bean's 'noble presence' dominated Wilkins's wartime experiences, and Wilkins later wrote that the direct influence Bean exerted on his work enabled him to look back on the war with something akin to gratitude.[13] It is also clear that Wilkins felt his involvement in the Historical Mission fulfilled a personal obligation to Bean and effectively completed his work as a war photographer.

It is frustrating that Wilkins left few insights into his wartime experiences, and although he has been the subject of three substantial biographies, attention has previously not been directed to his time working for the AWRS.[14] Nor does the archival record offer much help: a handful of letters

Photographer unknown
Informal portrait of Captain George Hubert Wilkins, c. 1918
Copy negative
Photograph reproduced courtesy of Ohio State University, Byrd Polar Research Centre Archival Program, Sir George Hubert Wilkins Papers
AWM P03171.003

THE MISSION

Wilkins with his assistant Joyce used the back of a British Mark V tank as a platform to record the advancing troops.

Unknown Australian official photographer
Captain George Hubert Wilkins MC (right) and Staff Sergeant William Joyce (left) record the 6th Brigade's advance to Montbrehain, 4 October 1918
Glass quarter-plate negative
AWM E03915

Attributed to Hubert WILKINS (1888–1958)
A group of men from the Australian 24th Battalion, commanded by Lieutenant Albert Sedgwick MC, wait in Elsa Trench for the artillery bombardment to lift before attacking Mont St Quentin, 1 September 1918
Glass half-plate negative
AWM E03142

scattered through the Memorial's official documents give only a meagre account of his years on the Western Front, and his papers in the Ohio State University have little relating to this period.[15] Until recently, the problem of understanding Wilkins's work as a war photographer has been exacerbated by a paucity of images that can be definitively attributed to him. When he worked as an official photographer during 1917 and 1918, images were not credited to individual photographers and no lists of 'who took what' appear to have survived. However, the Memorial is slowly unravelling Wilkins's personal history as it relates to his war work, tracking his and other official photographers' daily movements across the battlefields in an effort to attribute images to specific photographers.

In contrast, Lambert's life and career as an artist are relatively well documented and publicly accessible in archives, books and exhibition catalogues.[16] Lambert's selection as the Historical Mission artist was his second opportunity to work as an official war artist. In 1918 he had been appointed by the Australian High Commission to make an artistic record of the Australian Light Horse operating in the Middle East. Lambert was an inspired choice to work with the Australian mounted troops. He was an excellent horseman and had a cosmopolitan background. Born in Russia in 1873 of American and English parents, he spent his early childhood in Germany and was schooled in Britain before migrating to Australia, where he worked on a relative's sheep station. After a few years there he started art lessons in Sydney; in 1900 he won the prestigious New South Wales Society of Artists Travelling Scholarship and this allowed him to travel to Europe to further his studies. By 1914 he had become as well known in London for his work as a society portrait painter as for his flamboyant and extroverted character.

In December 1915 he tried to enlist in the British army, preferring either the Royal Engineers or Royal Artillery, but his application was declined, probably owing to his age.[17] He worked for a while managing forestry production for the British government before his appointment as an Australian official war artist in 1918 allowed him to fulfil his ambition of using his artistic skills as the best way to serve his country.

WILL DYSON (1880–1938)
Captain Wilkins operating a cinema machine on the Hindenburg Line, France, October 1918
Black crayon and pencil on paper, 43.2 × 29.8 cm
AWM ART09914

Will Dyson gained approval to go to the Western Front in December 1916 and worked there sketching events around him for some months before being appointed Australia's first official war artist in 1917. Dyson and Hubert Wilkins frequently moved around the trenches and front lines together and in October 1918 were both close to the Australian forward positions along the Hindenburg Line.

'Mud and mud' notes the 18th Battalion war diary on 8 November 1917: 'men cannot stand still long in one place without sinking up to their knees'.

Herbert BALDWIN (1880–1920)
Charles Bean struggles through the mud of Gird Trench in the Gueudecourt sector, France, late 1916
Glass half-plate negative
AWM E00572

THE MISSION

Attributed to Hubert WILKINS (1888–1958)
Men from the Australian 24th Battalion occupy a trench in front of De Knoet Farm on Broodseinde Ridge, 5 October 1917
Glass half-plate negative
AWM E01831

Oswald ('Ossie') COULSON (1872–1945)
Captain George W. Lambert, Official War Artist, sketching at Tiberias, May 1918
Glass half-plate negative
AWM B03210

Lambert had great respect for the Australian troops and considered his work as a war artist the best way to serve his country, recording military achievements 'for the sake of the brave men of all ranks out there'.

George LAMBERT (1873–1930)
Australian Light Horseman [alternative title **Digger**], Abbassia, Egypt, 1918
Pencil on paper, 31.6 x 24 cm
AWM ART02795

In 1915 Chauvel commanded the 1st Light Horse Brigade on Gallipoli. After the evacuation he was given command of the Australian and New Zealand Mounted Division and from April 1917 he commanded the whole Desert Column, later known as the Desert Mounted Corps.

George LAMBERT (1873–1930)
Lieutenant General Sir Harry Chauvel,
Egypt, 15 February 1918
Pencil on paper, 35.4 x 24.9 cm
AWM ART02734

THE MISSION

Lambert considered this little painting his 'most sincere and successful effort. It was painted in [the] heat of the day, a breathless heat, that made the camels lie down and look even more sorry for themselves than usual.'

George LAMBERT (1873–1930)
Wadi bed between El Arish and Magdhaba, Northern Sinai, mid-March 1918
Oil on maple wood panel, 19 x 24 cm
AWM ART02679

Lambert stayed at El Arish, south-west of Gaza, in mid-March 1918 and found the whitewashed buildings, bleached sky and surrounding desert an interesting aesthetic challenge.

George LAMBERT (1873–1930)
El Arish, Northern Sinai, c. 23 March 1918
Oil on wood panel, 15.3 x 22 cm
AWM ART02677

During his first tour as a war artist in the Middle East in 1918, Lambert studied the ground carefully, making oil and watercolour studies of places that became significant to Australian military history.

George LAMBERT (1873–1930)
Romani, Mount Royston in background,
Northern Sinai, 18 March 1918
Oil with pencil on wood panel, 22.2 x 30.6 cm
AWM ART02704

Lambert revelled in the light, landscape and atmosphere of the Middle East. Everywhere he looked there were 'glorious pictures' and he was surrounded by 'magnificent men & real top hole Australian horses'.[18] The small oil sketches he produced in places such as El Arish and Romani reveal his intense fascination with the desert environment and the work of the Australian soldiers. It was evident that he felt humbled by the dedication of his companions: 'Even if I had been inclined to slack myself, the example of the men who never tired forced me to be up early and paint and draw all day.'[19]

He travelled with the Australian troops through Egypt, Palestine and the Sinai for five months, and at the end of this tour he delivered 49 small oil paintings and nearly a hundred drawings. The work was considered an outstanding success and he secured an immediate commission to paint a major battle picture of the famous Australian Light Horse charge at Beersheba on 31 October 1917. Like Wilkins, Lambert was thoroughly committed to his work and supported the efforts of Bean and the AWRS to commemorate the service of Australian soldiers through the creation of the National Collection.

After seeing the paintings and drawings produced by Lambert during his time with the Light Horse, Bean was anxious for him to join the Historical Mission. Bean was in London in November 1918 and eagerly approached Lambert with the offer of the tour and a commission for two major Gallipoli paintings that would be central to the future museum displays.[20] These paintings were to represent two of the most significant events of the campaign: the ANZAC landing of 25 April 1915 and the charge of the 3rd Australian Light Horse Brigade at the Nek on 7 August 1915. The two subjects had been identified as essential for the national collection of war pictures and Lambert was considered the best artist to paint them. Lambert was not a radical modern artist and was, in Bean's words, 'transparently honest, [and] devoted, as a religion, to truth as he saw it'.[21] Bean recognised in Lambert a brilliant technician and a masterly storyteller, someone who would deliver work entirely compatible with his own standards of truth required in recording the war.

Both Lambert and Wilkins worked under the official commissioning scheme for artists and photographers to create their visual record of the Gallipoli site. However, Bean also wanted other men with him who could take care of logistics, travel arrangements and correspondence. For his personal assistant he chose Staff Sergeant Arthur Bazley, who had been with him most of the time since their departure from Sydney in late 1914.[22] Bazley had worked with Bean on Gallipoli and the Western Front before accepting a transfer to the newly formed AWRS in London in June 1917. Lieutenant John Balfour was Bean's other assistant.[23] He had served on Gallipoli as a staff clerk in the 1st Australian Division Headquarters and moved across to the AWRS in June 1918 to work as Treloar's assistant. During the Historical Mission Balfour managed the travel arrangements, organised the details of the camp on Gallipoli, and supervised the collection of relics in the field.

Resolving details of the ANZAC landing was paramount, so Bean also requested the help of an Australian soldier who had been part of the dawn landings. Although the war was over, the fulfillment of this request depended on the availability of men who could be spared from duty. He was assigned Lieutenant Hedley Howe, a young Victorian who had served with the 11th Battalion (a predominantly Western Australian unit) when they landed in the early morning of 25 April 1915.[24]

The Historical Mission's complement was rounded out with the nomination of two draughtsmen who would help in the ground survey and map-making. These were Lieutenant Herbert Buchanan and Sergeant George Rogers (another Gallipoli veteran), both of whom were engineers.[25]

MISSION OBJECTIVES

There were three key objectives for the Australian Historical Mission. Bean had been asked by the Australian government to provide a thorough report about the state of the Australian war graves and cemeteries that had been in Turkish hands since December 1915. This was an important and highly sensitive task that needed a quick response. Anxiety about abandoning Australian dead had increased greatly as the evacuation date approached. In the last few days there had been much activity to document the cemetery sites, and soldiers had tenderly repaired individual graves and tried to

From left to right: Captain Henry Casimir Smart, AIF Headquarters; Brigadier General Walter Ramsay McNicoll, commanding officer of the 10th Brigade; and Charles Bean.

Unknown Australian official photographer
Visiting the 39th Battalion in the front-line trenches at Houplines, France, December 1916
Glass half-plate negative
AWM E00087

Reclining on a piece of elephant-iron, Charles Bean watches the 2nd Division attack Malt Trench from a position near Martinpuich, France.

Herbert BALDWIN (1880–1920)
Charles Bean
26 February 1917
Glass half-plate negative
AWM E00246

provide more permanent headstones by replacing biscuit-tin name plates with solid wooden ones. Leaving Australians buried in foreign soil was, as Bean later commented, 'a cause of deeper regret to the troops and their people at home than any other implication in the abandonment of the Peninsula'.[26] Furthermore, unease in Australia had been heightened during the intervening years by fear that the grave-sites might be desecrated. The Armistice of Mudros, signed on 30 October 1918, ended hostilities with Turkish forces and allied troops quickly re-occupied the peninsula. An Australian-led group attached to the ANZAC Section of the British Graves Registration Unit was working on the peninsula by 10 November 1918.[27] Its first report of systematic desecration of graves caused great concern in Australia and led to the request for Bean to report as soon as possible on the state of the graves.

The second objective of the Historical Mission was to gather information for the proposed official history of Australia's involvement in the Gallipoli campaign. Evidence was Bean's primary concern. Material evidence – such as trench systems, cartridge casings, soldiers' kit and the bodily remains of Australian soldiers isolated from the main force – could be found in the traces of the fight that remained on and in the ground. These traces could provide important clues about the battles. However, seeing the place again, and having the time to investigate the relationship of the terrain to the military events, was also crucial to his work. Resolving details of the ANZAC landing was paramount, which is why Howe, a veteran of the landing, joined the group. Bean also proposed that draughtsmen and map-makers from the AWRS should accompany him to make precise surveys and to start preparing the maps and diagrams that would eventually be included in the volumes of the official history. Ever thorough, Bean believed it was equally important to know what the enemy had seen and to place himself in their position so that he might better understand their movements and motivations.

The identification of relics and the creation of a pictorial record was the third objective for the Historical Mission. The concept for the future war museum was based on the three types of records identified by Bean in 1917: written sources, visual records and physical objects. The power of the real object, together with the visual testimony of artists and photographers, would convey what words could not. Museum visitors would not only be able to read about the war, they would be presented with a rich and comprehensive visual experience. For this to be successful, it was essential that the most pertinent and evocative objects, photographs and works of art relating to Gallipoli be brought together. In particular, the paintings and photographs made in 1919 would record all the places where Australians had fought, providing a visual catalogue of the landscape. This catalogue would serve as a kind of atlas, enabling Australians to locate names and places connected to the momentous events of 1915 within specific landscapes.

These goals governed the work of the Historical Mission, and the principal participants were keenly conscious of the historical importance of the work about to be undertaken. Despite the enormous loss of human life on the Western Front, Gallipoli remained a landmark event in the Australian experience of war and this was the first opportunity for Australians to return to the site. Using different mediums, the historian, the photographer and the artist would each help to make the landscape and events visible to the Australian people and consequently help shape and inform the public history of the Gallipoli campaign.

William SWANSTON (1881–unknown)
and William JAMES (1887–1972)
The Golden Horn, from Galata Tower, Constantinople, 30 January 1919
Glass 10" x 12" negative
AWM G01783

CHAPTER THREE

TRAVELLING EAST

THE MONTH LEADING UP TO THE DEPARTURE OF THE Australian Historical Mission was hectic for Charles Bean. As well as preparing for the journey, he needed to record details of the battles in which the Australian soldiers had been involved before they had left the line in early October 1918. Hostilities on the Western Front had ceased on 11 November, and the Australian battalions were beginning to break up and depart from the European front lines. Having set a goal of trying to see two Australian battalions a day, Bean spent long days and evenings gathering information and interviewing men.[1] Additionally, he met with Australian official war artists and photographers to provide direction about appropriate subjects and made frequent visits to important battlefield sites to make detailed notes. One of these trips was to Péronne, in northern France, where he celebrated Christmas Day 1918 in the ruins, sharing a meal with Australian official war artist Lieutenant A. Henry Fullwood and assistant official photographer Sergeant William Joyce.[2]

While working across the battlefields on the Continent, Bean continued to put things in place for the Historical Mission. Since the initial approach to George Lambert, contract negotiations with the artist had not gone smoothly. Lambert was deeply involved in work on the large painting of the Australian Light Horse charge at Beersheba and was reluctant to delay this in order to go to Gallipoli. He was also not convinced that going there in the middle of winter would be good for his health (he was recuperating from malaria); nor did he think this would be the appropriate time for doing his preparation for the large canvases that might follow. On 11 December 1918 Lambert wrote to Captain H.C. Smart of the Australian High Commission indicating that he was 'very keen to paint a lasting memorial of the Charge of the 10th L H for example but [he] would need to be on the ground at the corresponding time of year.'[3] If he was going to paint the all-important Gallipoli pictures for the nation, he was adamant that his visit to the battlefields should coincide with the time of the original spring landing of ANZAC troops so that he could paint in the right atmosphere and light.

With Lambert uncommitted, Smart suggested other artists as potential replacements. But Bean stood firm: 'I am particularly anxious for Lambert to be the one to paint these special pictures.'[4] He tempted Lambert with the promise of a short stay on Gallipoli – 'two or three days so as to get the actual position' – and then a month or two painting in Egypt.[5] Lambert made a counter-offer: the rank of major, £2 per day in expenses and a widow's pension for his wife Amy in the event of his death. Just after Christmas the deal was closed. Lambert's rank was increased from honorary lieutenant to captain and he was granted travelling expenses of £1 a day. Should he die, however, Amy would not be entitled to a pension.

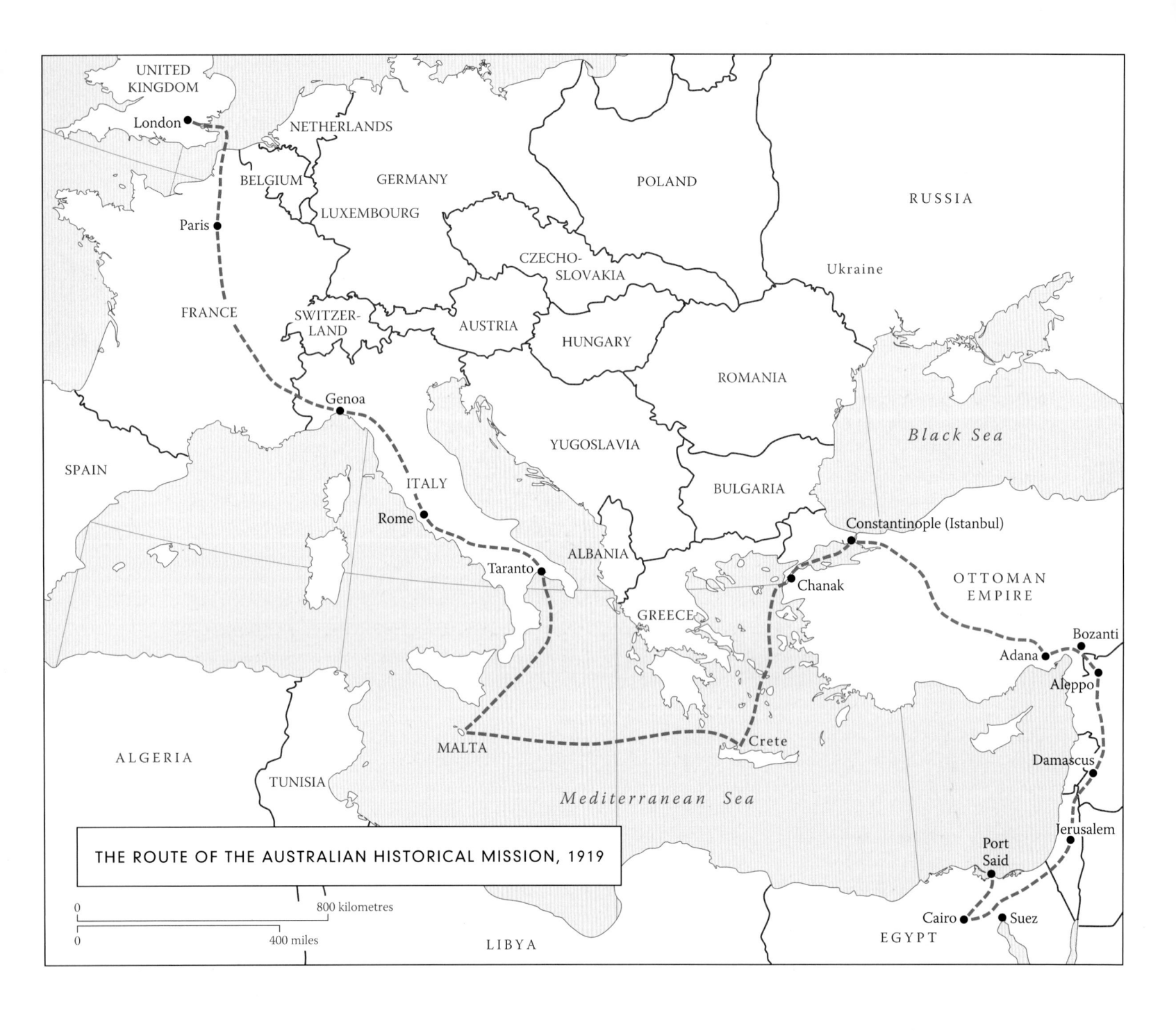

THE ROUTE OF THE AUSTRALIAN HISTORICAL MISSION, 1919

By early January, the Historical Mission personnel had begun to gather in London. Lambert packed up his work on the Beersheba painting; Bean, Hubert Wilkins and Hedley Howe came across from France; Bean's two assistants, John Balfour and Arthur Bazley, collected the diaries, notes and other material that would be needed in the field. The mapmakers, Herbert Buchanan and George Rogers, rounded out the Historical Mission. Rogers recalled that they all travelled on civilian passports so as to not contravene the terms of the Armistice.[6]

On Saturday 18 January 1919, the eight members of the Historical Mission assembled for the first time as they waited for the train that would take them down to Southampton for the Channel crossing. Some had not worked with each other before, but Bean was the glue that bound them together. He had played a large part in establishing and orchestrating the AWRS program; apart from directing the work of the official war photographers and artists, he had a clear vision of the future war museum, its collections, and its possible displays.

Of the three principal members of the group, Lambert, aged 46, was the eldest, but he clearly revelled in the discipline of military life and gladly took his direction from the 39-year-old Bean, whom he called 'Skipper'. Bean, with his red hair, blue eyes and glasses, was lean and active after four years of military life. He was usually noted as being shy and reserved, a dramatic contrast to Lambert, who was outrageously outspoken and extroverted. Sporting the 'golden beard, the hat, the cloak, the spurs, the gait, the laugh and the conviviality of a cavalier', he entertained everyone during the expedition with amusing yarns and displays of mimicry, often at the expense of stuffy British officers he had met.[7] Underneath the carefully constructed public persona, however, he was honest and devoted to his work, deeply sensitive, and prone to bouts of melancholy. Although he had been initially reluctant to head off overseas again so soon after his first stint as an official war artist, the promise of a commission for two major Gallipoli paintings proved irresistible. This second assignment gave him a further opportunity to record in paint all the traits and achievements of the Australian light horsemen he had come to admire so much in Palestine.

There is no record of Lambert and Wilkins meeting prior to 1919 but the two men quickly developed a lasting friendship. They were both practical men who shared a love of rural Australia, horses and outdoor life. Both were interested in mechanics and engineering, singing and fine music, and were well travelled and established in their chosen professions. Wilkins was engrossed in science and philosophy and could be drawn into serious discussion on almost any topic. He was said to be 'a man who rarely slept more than four hours out of the twenty-four', someone who was charismatic and had a great sense of humour.[8] Bean called him silent and watchful, a man who 'invariably sought his goal through adventure – the more dangerous the more acceptable'.[9] He had a deep respect for Wilkins and clearly considered him the best photographer for the tour, but there was another reason for his selection. Experienced in the rigours of military life, Wilkins had done outstanding work on the Western Front, consistently exposing himself to danger to obtain the records. Bean noted that the Gallipoli trip was 'partly … a reward for [Wilkins's] work' over the previous two years.[10]

Bean had carefully hand-picked the men for the Historical Mission. Howe, Balfour, Bazley and Rogers had been on Gallipoli during 1915 and the others had extensive wartime experience on either the Western Front or in Palestine. Bean could confidently expect that the members of his party, used to privation, danger and travelling in difficult conditions, would work together as a team when the Historical Mission got underway. However, a commitment to the exacting standards set by Bean and to the aims of the AWRS was as important to the success of the Historical Mission as familiarity with military life and the demands of travel. This sense of a common purpose was also expressed in the egalitarian travel arrangements for the group: 'we travelled together, lodged together, and ate together throughout our journey'.[11]

THE JOURNEY ACROSS EUROPE

With the party assembled, the Australian Historical Mission set out. Their plan was to head to Constantinople and then down to Gallipoli for three to four weeks' work on the battlefields. Uncharacteristically, Bean, who was exhausted from

From left to right: Lieutenant F.W. Eggleston, assistant to Sir Robert Garran; Lieutenant Commander J.G. Latham, Naval Staff Officer; Miss Eleanor Maida Carter, typist; Right Honourable Sir Joseph Cook, Minister for Navy, Australia; R. Mungovan, private secretary to Sir Joseph Cook; Right Honourable W.M. Hughes, Prime Minister of Australia; Captain H.S. Gullett, press secretary; Mr W.E. Corrigan, messenger; Sir Robert Garran, Solicitor General, Australia; Miss Wood, typist; and Lieutenant P.E. Deane, private secretary to Mr Hughes.

Photographer unknown
The Australian representatives at the Paris Peace Conference, 1919
Glass half-plate copy negative
AWM A02615

the last few weeks of frantic organising, kept only minimal notes as they made their way towards Gallipoli. However, Lambert kept a detailed journal and Balfour maintained a time-sheet of arrivals, departures and journey times. These documents, as well as Bean's own account of the journey written up much later, in 1948, give an introduction to the Historical Mission's progress across southern Europe.

The Channel crossing on 19 January 1919 was comfortable and uneventful, a marked contrast to the unease that had accompanied sea travel for the previous four years. After an extended period of always being close to the fighting, Bean had initially found peace bewildering, commenting that 'one doesn't get used to peace in a day'.[12] But, on a lighter note, he said that with peace he found 'the difference to one's personal comfort was as great as that between drinking castor oil and enjoying a cup of French chocolate'.[13] Rogers also found that after his experience as a soldier 'it was a strange feeling to be treated as a human being and stay in good hotels and have money to spend if necessary'.[14]

The next day they stopped in Paris where the Peace Conference had just opened at the Palace of Versailles.[15] Lambert noted that the Historical Mission met up with some of the Australian representatives, including the Prime Minister Billy Hughes and his press liaison officer Henry Gullett, both of whom were optimistic about the outcome of the conference.[16] Bean knew these men and considered them good friends. He had escorted Hughes around the battlefields during the war and had communicated frequently with him over the establishment of the AWRS and the future national war museum. After a short stint replacing Bean as official correspondent on the Western Front, Gullett had been appointed in October 1917 to head the newly established AWRS office in Cairo. Shortly afterwards, he was asked to write the official history of the AIF in Sinai and Palestine.[17] As well as briefing Bean on the conference, he discussed the work of the AWRS in Cairo,

and even suggested that Wilkins might take photographs in Palestine when the work of the Historical Mission concluded.[18]

In Paris, amid the excitement of the peace negotiations, the oppression of the war years gradually lifted from the group and their journey was subsequently punctuated with cultural excursions. In Paris they went to the Folies Bergère music hall – 'as harmless as a Sunday-school treat' – and in Rome took in a little sightseeing and the first night of the opera *Carmen*.[19] During their brief stop in Rome, some luggage was stolen, but fortunately Bean was carrying copies of some of his precious ANZAC personal diaries, the loss of which proved to be only a slight inconvenience. More seriously, Bean's trusted friend and assistant Bazley became gravely ill with the pneumonic influenza then raging at epidemic proportions in Europe. Bazley was hospitalised and forced to withdraw from the Historical Mission. He later rejoined the group in Cairo after the Gallipoli expedition.

The Historical Mission arrived at Taranto in southern Italy on 25 January. At once, Lambert started work, painting one small watercolour of the military rest camp where they stayed.[20] From Taranto they had a rough crossing to Valletta, the capital of Malta. Bean likened the pitching and shuddering of the ship to a dog shaking itself when wet, and most members of the Historical Mission, although seasoned travellers, became violently seasick.[21] They landed at Valletta and after recovering went to another opera, *Fedora* by Umberto Giordano.[22] Over the next two days, they visited the graves of Australian war dead. Australian wounded had been sent to the British base at Malta shortly after the Gallipoli landing, and there were about 200 Australian graves carved into the soft rock of Malta. The weather during their stay was poor, and Lambert was only able to start one oil sketch on cardboard of the ancient city overshadowed by a leaden grey sky. As they left Italy, the attention of the artist and the rest of the group became firmly fixed on their destination, Gallipoli.

Breaking their journey at the southern Italian town of Taranto, the Australian Historical Mission stayed two days at a military rest camp a few kilometres outside town. Bean described it as dreary and 'almost a replica of the bleak, muddy camps we knew so well in France'.

George LAMBERT (1873–1930)
Rest Camp Taranto, Italy, 25–26 January 1919
Watercolour, pencil and carbon
pencil on paper, 19 x 30.4 cm
AWM ART11393.358

TRAVELLING EAST

The Australian Historical Mission sailed through rough seas to reach Malta. The ship shook and shuddered; even Wilkins, an experienced sailor who had journeyed around the Arctic in wild weather, was seasick.

George LAMBERT (1873–1930)
Storm effect, Malta, Malta, 28–29 January 1919
Oil on canvas on cardboard, 20.1 x 30.2 cm
AWM ART02847

Charles BEAN (1879–1968)
Lemnos Island, 18 April 1915
Nitrate negative
AWM G00881A

The first time Bean visited Lemnos was in 1915. He had watched the troop transport ships gathering in Mudros Harbour in readiness for the landings on Gallipoli and wrote that 'the scene at night was studded with their lights'.

Charles BEAN (1879–1968)
Mudros Cemetery, 5 February 1919
Nitrate negative
AWM G02073

The Mudros cemeteries on Lemnos contain the graves of over 1,200 allied soldiers who were killed in the Gallipoli campaign.

While crossing the Aegean, Bean prepared everyone by presenting a two-hour lecture outlining the history of the ANZAC campaign. Lambert noted that the talk was 'thorough, interesting & highly instructive' and good preparation for his coming work.[23] They stopped off at Lemnos, the Greek island that had served as a base for the campaign and had been the site of hospital facilities for Australian casualties. While Lambert painted, Bean took the others to places of interest, including the cemeteries that held Australian war dead.

Lemnos was a significant transit point for the beginning and sometimes the end of countless Australians' involvement in the Gallipoli campaign. From here on, the Historical Mission's journey would start to intersect with those made by other Australians in 1915. They were reminded just how close they were to Gallipoli when they noticed the faint outline of the hilltops of the Dardanelles in the far distance. Bean wrote a poignant letter to his brother: 'Do you know, it seems funny to confess it, but I am as homesick as can be for Anzac.'[24]

On 6 February, as they sailed close to the southern tip of the Gallipoli peninsula, memories of the war were aroused. Rounding Cape Helles they saw the wreck of the transport ship *River Clyde*, grounded on 25 April 1915 to land British troops. The landing forces here had suffered terrible casualties. As the Historical Mission sailed past the cape, for the first time they could see the Turkish forts that guarded the wide entrance to the Dardanelles. The eastern and previously unknowable side of the peninsula held by Turkish forces throughout 1915 was now clearly visible across the narrowing stretch of water. Bean and Wilkins both took photographs of this first view of the forts across the straits and the gently rising outline of the Kilid Bahr plateau. Around mid-morning the Historical Mission landed at Chanak (now Çannakale), where most of the group disembarked to arrange the equipment, supplies, men and horses that would be transported to the base on the Gallipoli peninsula.

George LAMBERT (1873–1930)
Sunset, Lemnos, Greece, 4 February 1919
Oil on artist's board, 20.1 x 30.2 cm
AWM ART02853

Charles BEAN (1879–1968)
View across the Dardanelles from Chanak, showing ships in the background, 6 February 1919
Nitrate negatives
AWM G02069P

Combined from three negatives (G02069A–C), this panoramic view across the Dardanelles towards the Gallipoli peninsula shows the busy military and commercial shipping activity in the narrow channel. The small jetty on the left is the same as the one that appears in Lambert's painting, *Dardanelles from Chanak, effects of blizzard on Gallipoli*.

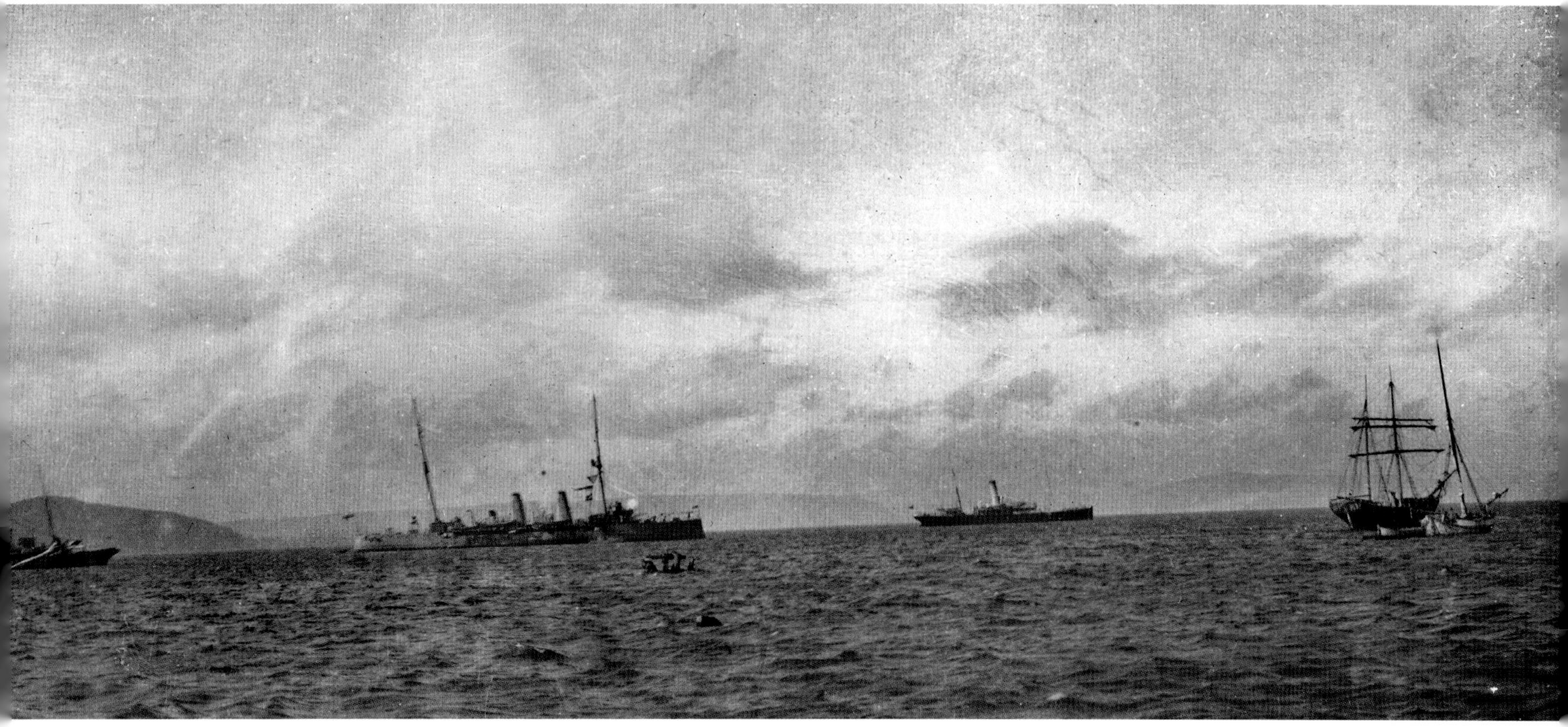

Photographer unknown
The Australian flag flying from the bow of HMAS *Parramatta* on her arrival in Constantinople harbour, 13 November 1918
Nitrate negative
AWM J03215

William SWANSTON (1881–unknown)
and William JAMES (1887–1972)
The forts at Kilid Bahr in the foreground, looking east across the Dardanelles to the town of Chanak,
January–April 1919
Glass 10" x 12" negative
AWM G01859

Photographer unknown (Photopress)
A street in the town of Chanak, showing the damage resulting from shelling by the British battleship *Queen Elizabeth*, c. 1915
Gelatin silver print
AWM H16637

Early in 1915, battleships such as the *Queen Elizabeth* replaced regular land-based artillery and pounded Turkish defences and the coastal towns.

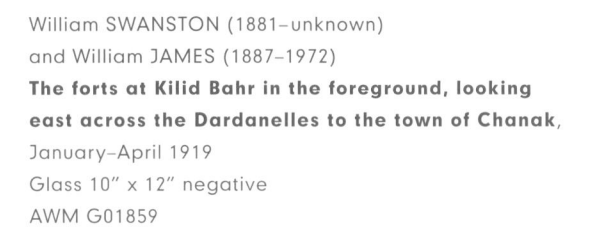

Bean and Wilkins continued on to Constantinople (now Istanbul), the capital of the Ottoman (Turkish) Empire. They travelled up the Dardanelles, all the time watching the shore and the outline of hills on the western side. The next day it was cold, wet and misty when they sighted the mosques and palaces of the capital. The already populous city was flooded with military personnel and displaced civilians who had taken up residence there while the terms of the peace continued to be negotiated following the Armistice. There was some suggestion that Wilkins could take a series of photographs of Australian destroyers and scenes of life aboard a troopship.[25] By the time Wilkins and Bean arrived in February, however, the Royal Australian Navy (RAN) ships that had been deployed for patrols and escort work in the region had departed just a month earlier.

Bean intended to spend a week interviewing staff at the headquarters of British forces stationed in Constantinople and meeting up with friends who might be able to help with the work of the Historical Mission. As he was intent on enhancing his understanding of military events by getting as complete a picture as possible, he also wanted to interview Turkish army staff to get their perspectives of the campaign. He prepared a list of over a hundred questions about troop deployments and tactics to ascertain what the Turks had understood and been able to see of the Australian forces during 1915. At the suggestion of British intelligence staff, the questions were forwarded to the Turkish authorities but with an additional request for the Historical Mission to have the assistance of a Turkish officer who had served on Gallipoli. It was hoped this officer would camp with the Australians 'on terms of mutual respect' and walk with them over the battlefields to provide a Turkish perspective. Turkish military authorities responded positively to this request and advised that they would nominate an appropriate officer to assist.

As the Australian Historical Mission passed Cape Helles and headed towards Chanak, Bean reached for his camera and took a hurried photograph of the Kilid Bahr plateau, which dominates the southern end of the Gallipoli peninsula.

Charles BEAN (1879–1968)
Kilid Bahr Plateau under cloud, taken from Chanak, 6 February 1919
Nitrate negative
AWM G02068

and looking at ancient architecture and Greek, Roman and Christian relics. They were keen to get back to Chanak, but snowstorms prevented their departure until 13 February.

The rest of the Historical Mission remained stranded in Chanak, the poor weather preventing them from crossing over to the Gallipoli peninsula. They were staying in what Lambert called a fifth-rate boarding house, which had neither coal nor wood for fires.[27] With blizzards and raging winds for much of the week, most of the group stayed indoors or attempted to make friends with some of the locals.

For his part, Lambert found other amusements. He sketched a local Turkish man and also joined in some of the regular entertainment such as a Gilbert and Sullivan performance put on by men of the British 28th Division, which was watched by Britons and Turks alike. Lambert was invited to dine with a British officer who described the artist as a:

> shaggy fellow with a jutting yellow beard and ultra virile affectations and also the rather silly affectation of knowing nothing about English manners ... Like most painters, he has a remarkable freedom of wit, even if he rather overdoes the part of the rude Australian.[28]

George LAMBERT (1873–1930)
Head of a Turk, Chanak, Turkey, 11 February 1919
Watercolour with pencil on paper, 33.4 x 28.3 cm
AWM ART02863

While waiting at Chanak, Lambert made this portrait study of a Turkish veteran who had fought against Australian troops in 1915.

When they had a moment to spare, Bean and Wilkins did some sightseeing. Wilkins, who had visited Turkey during the First Balkan War in 1912, had friends in the city who received them warmly. It was Bean's first visit to Constantinople and his studies in history and the classics hardly prepared him for the majesty of the buildings and the depth of culture in the city he described as the 'centre of Western civilisation'.[26] The two men spent their time visiting mosques and museums,

Photographer unknown
View of Constantinople harbour, 5 November 1919
Glass half-plate copy negative
AWM A02713

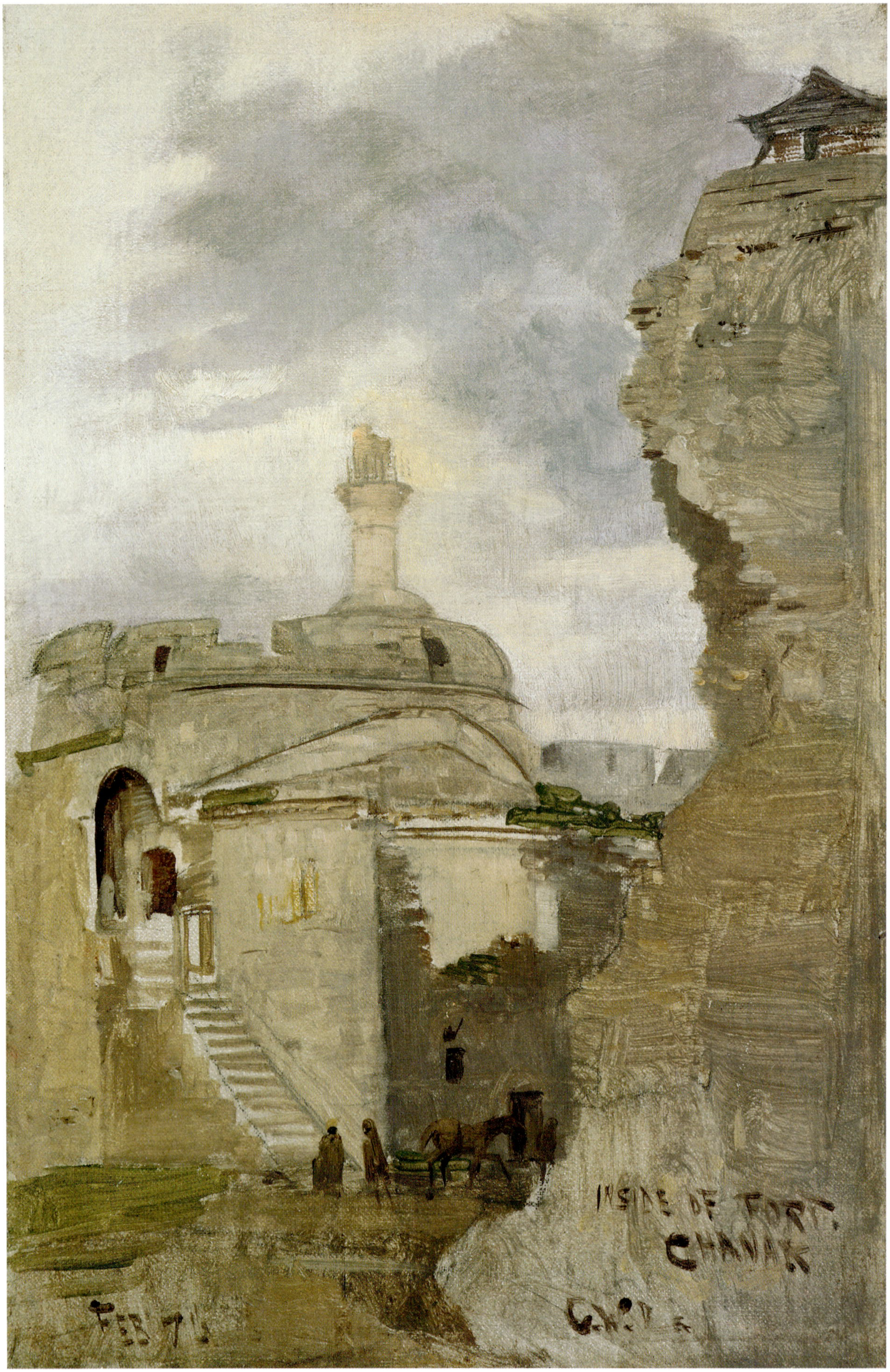
INSIDE OF FORT.
CHANAK
FEB 7'

He obviously found some of Lambert's showmanship irritating but liked the artist and spent time with him, watching him paint and conversing on such varied topics as art, wild animals and boxing.

In the week that Lambert was at Chanak, he painted and sketched his surroundings. In these few works we can see him becoming focused on his task of making an artistic record of sites and scenes related to the ANZAC story. His practice, as it had been in Palestine, was to make oil sketches on small prepared artist's boards and wood panels. Sometimes he quickly roughed in the outlines of the composition with a soft pencil and then blocked in large areas before reworking the fine detail. The fine-grained panels were honey coloured, and Lambert often deliberately left areas unpainted, the bare wood becoming, in effect, another colour in his palette.

With blizzards and rain on many days during his stay at Chanak, he had to work quickly and under difficult conditions. The German commander of the Turkish forces during 1915, Liman von Sanders, had used Chanak as his headquarters and evidence of the war was scattered through the small seaside town. Lambert concentrated his work around the shoreline that looked across the Narrows to the peninsula. He made a delicate study of an ancient fort, Çīmenlik, damaged by British naval bombardments, and another of Turkish guns guarding the entrance to the Dardanelles.

While the main focus of *Big gun emplacement, Fort of Chanak* is the artillery and fortifications, the outline of the peninsula looms in the background.

The dominant presence of the peninsula and the anticipation of getting there is most evident in two oil studies where Lambert has painted the peninsula as seen from the other side of the Narrows. *Dardanelles from Chanak, effects of blizzard on Gallipoli* was painted on a dull, stormy day. Broad and energetic brushwork creates the feeling of wild, wintry weather and all but the barest details of the opposite shore of the Dardanelles is obscured by a blanket of snow. The second sketch, *Gallipoli, from the Chanak side*, was painted on a sparkling but windy day. Lambert chose a wood panel for this lively sketch, leaving areas of the warm wood colour as highlights and clearly showing the distinctive outlines of the Kilid Bahr Plateau.

Despite the frustration of being stranded, Lambert was relatively pleased with the work he had done at Chanak: 'I have managed to make four sketches, quite passable, of points of interest but we look ever towards Anzac & I am glad to say we start for Kilid Bahr the fort on the Peninsular [*sic*] side today & go into camp on the ground.'[29] With the weather abating and Bean and Wilkins about to join them from Constantinople, the Historical Mission prepared to make the crossing to Gallipoli to begin the main work of the expedition.

George LAMBERT (1873–1930)
Inside the fort, Chanak, Turkey, 7 February 1919
Oil on canvas on wood panel, 30.4 x 20.2 cm
AWM ART02832

On the Asian side of the Dardanelles at Chanak stands Fort Çimenlik, one of a chain of forts and gun emplacements that opposed the British and French navies attempting to force their way through the narrows towards Constantinople in 1915.

George LAMBERT (1873–1930)
West Mudros rest camp, Lemnos,
Greece, 4 February 1919
Oil on canvas on wood panel, 20.9 x 31.7 cm
AWM ART02823

George LAMBERT (1873–1930)
Big gun emplacement, Fort of Chanak,
Turkey, 6 February 1919
Oil on artist's board, 20 x 30.2 cm
AWM ART02835

While Bean and Wilkins went on to Constantinople, Lambert and the rest of the party stayed at Chanak. A week-long blizzard kept them from crossing to the Gallipoli peninsula and Lambert complained, 'Snow blizzards ice & general discomfort. No coal or wood and a damp gloomy fifth rate house called the Lion Hotel, may I live to forget it.'

George LAMBERT (1873–1930)
Dardanelles from Chanak, effects of blizzard on Gallipoli, Turkey, 7–14 February 1919
Oil on artist's board, 20.1 x 30.1 cm
AWM ART02833

TRAVELLING EAST

Despite the bitterly cold weather and snowstorms, Lambert made four paintings and some sketches while he was stranded at Chanak. But his mind was on Gallipoli and he wrote on 14 February 1919, 'We look ever towards Anzac & I am glad to say we start for Kilid Bahr the fort on the Peninsula side today & go into camp on the ground.'

George LAMBERT (1873–1930)
Gallipoli, from the Chanak side,
Turkey, 7–14 February 1919
Oil on wood panel, 23.7 x 35.6 cm
AWM ART02858

This oil was painted in Legge Valley where the Australian Historical Mission camped. Lambert found the landscape 'dull but interesting as a problem in monotonous colour' and it reminded him of 'moorland in Yorkshire or South England'.

George LAMBERT (1873–1930)
Behind the Turkish lines, Gallipoli,
Turkey, 18–21 February 1919
Oil on wood panel, 13.6 x 22.2 cm
AWM ART02829

CHAPTER FOUR

LAMBERT
SETS TO WORK

ON 14 FEBRUARY 1919 THE HISTORICAL MISSION BOARDED a steel motor lighter (or 'beetle') and started to cross the narrow sea passage to the Gallipoli peninsula.[1] They took with them a wagon-load of equipment, including camera gear and painting supplies, five horses and four mules. Accompanied by a work party of eight British soldiers from the 28th Infantry Division based at Chanak, the men of the Historical Mission approached the peninsula from the Asiatic side of the Narrows. The view of the land looming before them was completely new to the veterans of the 1915 campaign – this was the military objective never seen and never achieved. The short journey brought them to the peninsula, and they were further reminded that they were now behind the 1915 enemy lines when they stayed for the night in an old hospital camp surrounded by about 3,000 Turkish graves. The next day they travelled towards the ANZAC area and set up camp in Legge Valley, immediately behind Lone Pine, where they were protected from the wind and the worst of the weather.

The temporary residence of the Historical Mission was close to the Australian section of the Graves Registration Unit, commanded by Captain Cyril Hughes.[2] Another group of men led by Lieutenant William James from the AWRS was also nearby. James and his group were already known to Charles Bean as they had met in Constantinople a week earlier when James was on short leave from Gallipoli.[3] James and Sergeant William Swanston, a photographer on secondment from the Royal Air Force (RAF), had arrived on Gallipoli on 30 December 1918 to undertake work on behalf of the AWRS branch in Cairo. They were directed to take record photographs of sites on the peninsula related to Australian actions and to identify potential objects for the museum collection.[4] They had already made a preliminary survey, working systematically back and forth across the battlefields collecting, tagging and describing relics. Many of the major battlegrounds had been photographed using a large-format glass-plate camera on loan from the RAF photographic section in Egypt. James and Swanston's work was seen as complementing that of the Historical Mission and the two groups often worked together on Gallipoli.

The Australians were only a small part of the fluctuating population on the peninsula in February 1919. Civilian visits to the area were restricted, but the Armistice signed with the Ottoman Empire only three months earlier had led to increased military activity in the region.[5] The terms granted Britain and her allies the right to occupy the forts that controlled the Dardanelles, and therefore the shipping routes up to Constantinople and through to the Black Sea. The Turkish troops that had garrisoned the peninsula since 1915 were quickly displaced when the British took control of the area. Many Turkish soldiers remained on the peninsula, however, with no apparent organisation to support them.

THE APPROACHES TO CONSTANTINOPLE, 1919

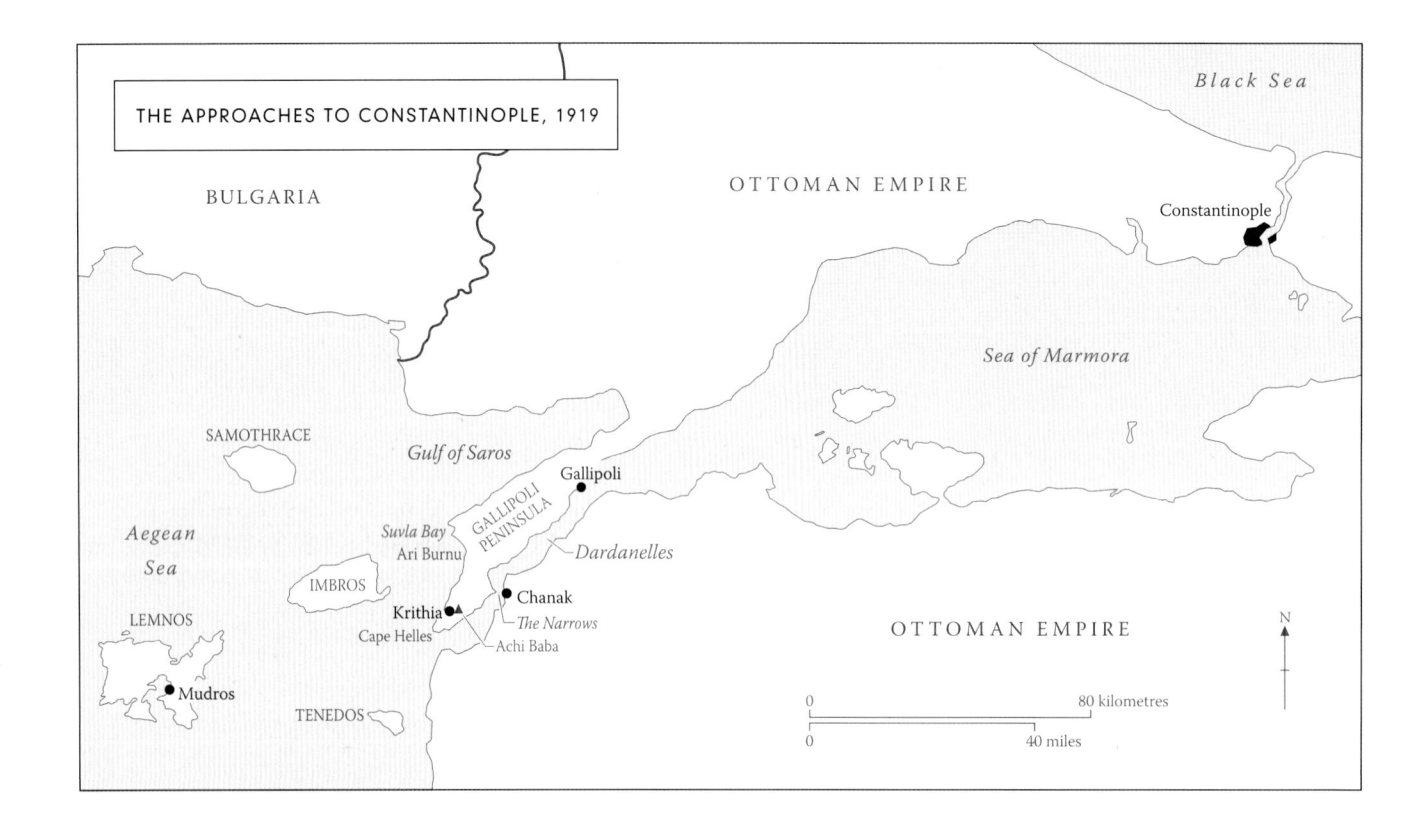

Their uniforms were ragged and they wandered the villages and various camps looking for work and food.[6] Amid it all, civilians were trying to remake their lives, repairing damage to land and buildings, and preparing the ground for spring crops. It was late winter, food was scarce, and the Australian camps were often visited by women and young children begging for food.

Hughes's Graves Registration Unit had been in the ANZAC area since November 1918 to attend to the Australian cemeteries.[7] Bean described Hughes as a 'tall, thin, breezy Tasmanian, country-bred and gifted with that easy confidence in tackling men and situations that seems to be more easily acquired in the Australian bush than in most environments'.[8] Hughes was a Gallipoli veteran and had secured another Tasmanian, Sapper Arthur Woolley, as his assistant.[9] Their job was to probe the ground, locate and identify the remains of Australian soldiers, rebury them and tidy the cemeteries in preparation for the civilian visitors who were expected to travel to the battlegrounds. With over a third of all Australian dead on Gallipoli having no known graves, the gruesome but essential work being undertaken by the Graves Registration Unit highlighted the tragedy of the campaign.

Australian numbers were boosted by a detachment of the 7th Australian Light Horse Regiment, which arrived on Gallipoli on 4 December 1918 as part of the forces taking control of the Dardanelles. They also took the opportunity to tour the battlefields in memory of their fallen comrades. Soldiers who had previously been at ANZAC had an intense interest in revisiting the battlefields, identifying old positions they had held, and 'dug-outs which had been their home for so long, and under such hard conditions'.[10] Equally, others 'were keen to see all the places whose names were familiar to them' from the accounts of Gallipoli veterans.[11] When the 7th Light Horse departed, several of the men remained behind to help with the various work parties operating across the peninsula.

Hubert WILKINS (1888–1958)
The Australian Historical Mission disembarking at Kilid Bahr, Chanak in the distance, 14 February 1919
Glass half-plate negative
AWM G01880

The 7th Light Horse Regiment was garrisoned on Gallipoli between 4 December 1918 and January 1919 and worked on salvaging and clearing material from the area as well as providing a strong military presence. Several members of the regiment were seconded to the Historical Mission and the AWRS parties to help with their work.

Photographer unknown
Unidentified members of the 7th Light Horse Regiment posing with one of the Krupp 14-inch guns mounted at the Turkish Hamidieh II Battery, south of Kilid Bahr, Gallipoli,
c. December 1918–January 1919
Copy negative
AWM P05460.003

The Australian Historical Mission was accompanied by soldiers from the British 28th Division as well as from the 7th Light Horse Regiment. The large tent served as a kitchen and sleeping area, while one of the bell tents was set up as a studio for George Lambert. He is shown reclining in front of the large mess tent on the left; behind him is John Balfour, followed by Herbert Buchanan, Charles Bean (seated), George Rogers and Hubert Wilkins.

Hubert WILKINS (1888–1958)
Members of the Australian Historical Mission in front of their camp, in Legge Valley behind Lone Pine, 14 February–10 March 1919
Glass whole-plate negative
AWM G02039

WALKING THE GROUND

The Historical Mission group occupied a large marquee that served as a kitchen as well as sleeping quarters for everyone. George Lambert had a separate bell tent to use as a studio when bad weather prevented his working in the field. Photographs indicate that there were a couple of other tents, probably used for preparing food and protecting essential supplies from the marauding wild dogs. A reliable water source was essential, and the group's camp was near one of the few wells in the area. Water was also necessary for developing negatives and making photographic prints.

The first few days established a pattern for the Historical Mission. Breakfast was at 8 am and Bean and his men moved off an hour later. If the day involved covering a major site such as ANZAC Cove, the Nek, Gaba Tepe or Hill 60, the entire group would walk the ground with those who had been there in 1915, sharing their stories. Lambert might break away to start a painting, and his assistant, Lance Corporal William Spruce of the 7th Light Horse, would help to carry the painting equipment in a large case nicknamed 'Lambert's coffin'.[12] Spruce managed the horses and a troublesome pack mule and occasionally stood in as a model for Lambert. The artist described Spruce as a 'dinkum Australian' who helped keep his temper in check.[13]

Hubert Wilkins and Bean generally worked together, criss-crossing an area, making notes, gathering evidence and taking photographs. On occasion they were joined by the AWRS men, James and Swanston, who took some direction from Bean about what to photograph but generally followed their own work program. Once an area was documented, John Balfour stepped in, tagging and collecting items suitable for the museum. Back at camp in the evenings, the men from the three groups often ate together and recounted the day's findings. The Historical Mission members continued their work into the night: Bean cross-checked and completed his notes, Wilkins developed his glass-plate negatives in a dark room made from a salvaged ship's tank, Herbert Buchanan and George Rogers worked on their maps, and Lambert painted in the bell tent until the light gave out. Balfour was responsible for the camp, maintaining supplies and documenting and packing objects for the museum collection. At the end of a long day there was usually 'five minutes of gossip' before they bunked down for the night.[14]

When Major Zeki, the Turkish officer who was assigned to help Bean with a Turkish perspective, arrived on 21 February, the Historical Mission's daily routine was intensified to make best use of his presence. Sensitive to cultural differences, Bean gave instructions that mess-food was not to contravene Muslim dietary requirements. Although neither Bean nor Zeki spoke each other's language, they could converse in French. Zeki also carried an English magazine with words and phrases that he understood marked so that he could point to them. During their long conversations Bean tried to write down as accurately as possible the Turk's responses. It seems remarkable that in early 1919, only a few months after the Armistice, the communication between former enemies could be so cordial, but the relationship was developed in a spirit of goodwill and cooperation from both sides.

The Turkish perspective was essential for Bean's work, and Zeki was well placed to provide a thorough and exact account. He had commanded the 1st Battalion of the 57th Turkish Regiment when the Australians landed and later served at German Officer's Trench, Lone Pine and Hill 60, where he commanded the 21st Regiment. A quietly spoken, reserved man, Zeki graciously submitted to Bean's extensive 'cross examinations' and later followed up with a written response to several of his questions.[15] In this, he detailed Turkish troop positions, command structure and specific details of military events all of which were important components of the story Bean was trying to understand. After a long day's work walking the battlefields and examining Turkish positions, the Australian historian and the Turkish staff officer would spend the evenings sitting in the mess-tent going over details of critical events from 1915.

On his arrival at ANZAC, Bean had been keen to revisit the places where he had spent many months hunkered down writing press despatches. He took Lambert and Wilkins straight down to ANZAC Cove. He showed them Shrapnel Gully, where men had lived like 'sand-martins' in their small dug-outs, and the beach that had been as busy as 'Broadway

Major Zeki had been assigned by the Turkish military authorities to help the Australian Historical Mission survey the Gallipoli battlefields. Zeki had also commanded Turkish troops opposing the landing of Australian and New Zealand troops in 1915 and then fought at Lone Pine, German Officers' Trench and Hill 60.

George LAMBERT (1873–1930)
Major Zeki Bey, Commandant of Turkish regiment at Gallipoli, Turkey, 28 February 1919
Pencil on paper, 30.6 x 23.4 cm
AWM ART02868

Lambert found it incredible that the soldiers had managed to scale the steep and crumbling slopes to establish a foothold on the peninsula. He revisited the Cove on 5 March 1919 to work on an oil painting, noting in his journal, 'I did a picture, not a sketch, of Anzac Cove, chiefly palette-knife, and quite like it.'

George LAMBERT (1873–1930)
ANZAC Cove, Gallipoli, Turkey, 5 March 1919
Oil with pencil on wood panel, 34.5 x 45.7 cm
AWM ART02839

This colour photograph was taken from a position close to where Lambert painted *ANZAC Cove*. The artist and photographer frequently worked alongside each other and even chose similar vantage points from which to produce their images.

Hubert WILKINS (1888–1958)
A view of ANZAC Cove as it appeared in February 1919
Paget colour plate
AWM P03631.232

Charles BEAN (1879–1968)
Hell Spit graveyard showing graves 'faked' (remade) by the Turks, c. 16 February 1919
Nitrate negative
AWM G02085

When he arrived in the ANZAC area, Bean made a tour of the cemeteries with Cyril Hughes from the Graves Registration Unit. Bean noted that the Turks had remade some of the graves to give the impression of their being well tended.

or Charing Cross'.[16] Turks had salvaged much of the portable war material abandoned after the evacuation, but on the beach there were still the remains of the concrete bases of water-condensing plants and in the shallows a stranded barge and two of the original lifeboats used at the landing. Apart from a new access road and barbed-wire defences constructed by the Turks, the area remained relatively unchanged. It was, as Wilkins described it, 'the most impressive battlefield I have seen'.[17] Yet Bean noted the subtle differences that now defined this as a postwar landscape: they rode horses through areas that in 1915 had been too dangerous to crawl across, and the complete silence was marked only by waves washing against the shore.

The day after camping in Legge Valley, Bean toured the ANZAC area with Hughes and then cabled his preliminary report on the condition of the Australian graves and cemetery sites to the Australian High Commission. Contrary to rumours, there had not been systematic desecration of the graves and only a few cemeteries had been interfered with. Most of the damage had probably been done by local inhabitants and soldiers looking for war loot or by dogs scavenging across the battlefields. In his report Bean commended the work of Hughes and Woolley in locating graves and encouraged the government to provide more resources to complete the task. He made a number of recommendations in his first report, including that the complete ANZAC site be vested in the Imperial War Graves Commission and that the small cemeteries and isolated graves remain where they were.[18] He made the point that Australians working with the Graves Registration Unit should focus on the ANZAC area and continue to supervise any work. Although the appearance of the battlefield should not be altered, he suggested, the cemeteries were a different matter and should be landscaped with Australian plants. Bean correctly anticipated that the site would be visited by Australians and he believed that the whole area should reassure visitors and families back in Australia that the remains of their dead were being carefully tended and watched over by fellow Australians.

FOLLOWING THE STEPS OF THE LANDING

The next day, 17 February, the entire group started to retrace the events of the landing and to walk what Lambert described as the most important ground. Bean, of course, could provide an eyewitness account of what he had seen from the ship and after he reached the shore, but he made certain that another person who had participated in the landing was there to lead the party through the events of 25 April. Lieutenant Hedley Vicars Howe had been in the first wave of troops to land before dawn on North Beach near Ari Burnu.[19] Howe had been working in the Western Australian pearling industry before enlisting and Bean described him as a fair-haired 'young scallywag' full of adventure.[20]

The ANZAC landing of 25 April 1915 had been promoted by the Australian press as something that would 'live for all time in the history of Australia; in the first place because it was an operation without parallel [and] in the second place because it marked the debut of Australasian troops upon the military stage of Europe'.[21] Furthermore,

Hubert WILKINS (1888–1958)
Looking towards Ari Burnu across the ridge which Lieutenant Howe climbed on landing,
17 February 1919
Glass whole-plate negative
AWM G01870

Hubert WILKINS (1888–1958)
The two winding tracks leading down from the crest on Plugge's Plateau which were used by the troops on the day of the landing, 17 February 1919
Glass half-plate negative
AWM G01881

Hubert WILKINS (1888–1958)
Looking from the crest of Plugge's Plateau to Russell's Top, which was first reached on the morning of the landing, across the Razorback to the Sphinx, 17 February 1919
Glass whole-plate negative
AWM G01871

On his first day walking the battlefields Lambert saw the Nek, which had been the site of a great tragedy for Australian troops on 7 August 1915. He thought the landscape was a 'wonderful setting for the tragedy', which he would later paint on a large canvas. In two hours he did what he called a 'Buckshee Souvenir' of the ground over which the Australians had charged.

George LAMBERT (1873–1930)
The Nek, Walker's Ridge, site of the charge of the Light Horse, Turkey, 17 February 1919
Oil with pencil on wood panel, 24.1 x 35.6 cm
AWM ART02856

these troops had accomplished a feat that had earned them 'imperishable fame'.[22] But the actual landing had been chaotic, with boats pulled or pushed out of their allocated positions and men losing contact with their officers. Trying to move through the rugged and difficult terrain had further broken up the order of the units. Some became isolated from the main force and were never heard of again, while others were blocked by the Turks and so could not reach their positions. At the end of the day, Australian and New Zealand troops were scattered across a large area, and had begun digging into positions that moved little over the next eight and a half months. But despite achieving popular acceptance as a defining moment in the nation's history, the events of the day were not known in sufficient detail for Bean to form an accurate military account. In the projected scheme for the war museum, the events of that first day ashore on Gallipoli would be a feature story told through objects and works of art.

Bean's method for his historical research was to accumulate individual stories that he could then use to corroborate others. He built a web of information, each new piece consolidating and extending his knowledge. The Historical Mission provided the opportunity to visit the site before it changed too much, so as to follow 'whatever traces the event had left there'.[23] Howe's story of the landing was an invaluable piece in the complicated puzzle, especially as it was told while they covered the actual ground where it had happened. Howe's memories provided the impetus for the work of the entire group as they traced his journey of nearly four years earlier.

As Howe guided them, Bean made notes and quick sketches that located events in the landscape and Balfour marked the location of objects they came across. Alongside them, Buchanan and Rogers also sketched out details that would later be used to make diagrams and maps for the official histories. Lambert started to identify sites from which he could paint his 'art records', and Wilkins took photographs that covered important aspects of the story.

Howe had a good memory and a 'gift for description' and led the Historical Mission to the exact spot where, nearly four years earlier, he had landed in the dark, thrown off his pack and started to climb the slopes. He showed them the beach where the boats had pulled into shore and the men had rushed forward to take shelter under small sandy cliffs. He then led them towards the heights known as Plugge's Plateau, where, in the semi-darkness, he had climbed the lower slopes. He related how he had seen men dig their rifle butts into the gravelly ground to get enough purchase to lever themselves forward, all the time fighting the dense prickly scrub of the hillside. He also described coming across two Turks, one of them dead, the other still conscious and clutching for a water bottle despite suffering from a terrible head wound. This was Howe's first sight of the enemy. Now, as they walked this ground, Wilkins stopped to take photographs of the terrain and the barbed wire still entangled in the undergrowth. He took photographs of specific Turkish trenches that Howe had overrun in 1915 and a view from one trench towards the shore to show what the Turks had seen of the approaching Australians.

After about 15 minutes of steep climbing, they came to the top of Plugge's Plateau, where Howe's unit had rushed across the open area to dislodge any remaining Turks. Not long after 7 am on the day of the landing, Bean had witnessed something of this event from the ship and recorded it in his diary: 'Our men seen on top of ridge … Men quite plainly visible in large numbers – entrenching slightly behind Hilltop – walking in quite uncovered'.[24] Continuing his story, Howe recalled that when he got to the top he saw his first exploding artillery shell: it 'burst in a fleecy white puff' over the scene.[25] From here he had a clear view across to the jutting peak of the Sphinx and down the steep spurs and gullies of the Sari Bair Range back to the beach.

LAMBERT'S MEMORY NOTES

After listening to Howe's account of the landing, the group walked along Russell's Top to the narrow piece of land called the Nek. It was here on 7 August 1915 at dawn that the 3rd Australian Light Horse Brigade had made what Bean described as 'the bravest charge in Australian history'.[26] Four successive lines of men from the 8th (Victorian) and 10th (Western Australian) regiments had run towards the Turkish trenches across the stretch of no man's land that, at its narrowest, was only 30 metres wide. Each line was met with 'a perfect sheet of flame' and cut down.[27] The losses were horrendous.

It had been decided that the charge at the Nek would be the subject of one of the major paintings that Lambert would produce for the future museum. Eager to commence his work, Lambert broke from the Historical Mission and late in the afternoon of 17 February made a quick oil study. He wrote that he was impatient and 'could not wait for the proper time which was just before sunrise'.[28] A fierce wind and limited time made him work quickly, and he only blocked in the spot's key features: the bare and broken earth, the closeness of the opposing trenches, the ground sloping away on either side, the distinctive outline of the distant ranges in the background. Bleached bones scattered across the ground reminded the artist of the 'terrible sacrifice' made at this place.[29]

He returned to the site two weeks later, arriving before sunrise to make sure that he got the right light effect. Although he was excited by the 'wonderful setting' that the landscape afforded for his painting, he found it cold, bleak and lonely: 'The Jackals, Damn them, were chorusing their hate, the bones showed up white even in the faint dawn and I felt rotten.'[30] *The Nek, Walker's Ridge*, with its hastily laid-in areas of ground, churned earth and sky, is a melancholy work, and this mood is accentuated by the figure of a lone soldier looking across the narrow field. At his feet, a skull and thighbone poke out of the earth. This is the only work Lambert did at the Nek, which is surprising considering the commission he had been promised.

Lambert approached his work with total dedication and commitment to achieving accuracy as well as artistry in his paintings. He understood that the large history paintings he would create back in the studio would probably be his most important public commissions. They would go straight into the museum collection, which would be pivotal to the way the nation commemorated and remembered the war, and would also provide an emotionally charged artistic interpretation of events. To be accepted by the veteran community and the general public, the paintings would have to represent events accurately and tell the grand as well as the individual stories in a way that engaged contemporary audiences.

Howe's personal story inspired Lambert, and many of its details would find their way into one of his large paintings, *ANZAC, the landing, 1915*. Additionally, walking the ground with Gallipoli veterans at places such as the Nek, and seeing evidence of the human tragedy, helped him to identify which components he wanted to convey in his paintings.

The small oil on wood panel sketches that the artist made on Gallipoli were not necessarily intended to be preliminary studies. In fact, very few of all the sketches Lambert made on his tours as a war artist were produced as literal studies for larger paintings. Instead they operated as 'memory notes', recording the colour, light and character of the landscape. He strove to paint at the same time as the given event, going out at dawn to see how the light played across the landscape, or, if required, painting at midday with the sun directly overhead. He also captured details of the Gallipoli landscape, such as the late winter plants and flowers, a distant blue outline of ridges, or the pock-marked battlefield in front of him. A few of these details found their way into later works, but the oil sketches Lambert made on the spot were essentially part of the process of getting the 'feel' of the place.

Working around the ANZAC area, Lambert painted two detailed views of the distinctive feature known as the Sphinx. *The Sphinx from Plugge's Plateau* was commenced on 18 February, a day after Howe recounted his experiences of the landing. Lambert took up a position on the edge of Plugge's Plateau, where he had a fine view across to the sharp features of the crumbling gravel ridge. This beautiful little painting, which captures the clarity and atmosphere of morning light across the rugged feature, shows the artist's complete control of his medium. Yet Lambert was constantly hampered by poor weather and needed three sessions to complete it. Biting winds, rain and sleet froze his fingers and made his work very difficult. He sat on the edge of a precipice waiting for the sun to shine so that he could quickly 'dash in' more of the image. Another sketch of the Sphinx was painted on a grey day when the dull light flattened out all the detail, leaving only a distinctive outline against the grey sky.

LAMBERT SETS TO WORK

81

The rugged sandy clay ridge that dominates the ANZAC Cove area was named the Sphinx as it reminded Australian soldiers of the ancient sculptures they had seen in Egypt prior to embarking for Gallipoli. Lambert went out at dawn to get the effect of the early light raking across the cliff faces in this detailed study of the Sphinx.

George LAMBERT (1873–1930)
The Sphinx from Plugge's Plateau,
Gallipoli, Turkey, 18 February and 4–5 March 1919
Oil with pencil on wood panel, 24.8 x 35.6 cm
AWM ART02846

Lambert believed it was important to make accurate studies of the local flowers and plants so as to get the authentic feel of the landscape. On 27 February, when it looked as if bad weather was setting in, he collected these wildflowers from Gaba Tepe. He spent most of the following two days in his tent-studio painting this still life. By the end of the second day he had finished it but was anxious to return to more military subjects; he wrote, 'I should be painting against time up at the Nek!'

George LAMBERT (1873–1930)
Gallipoli wild flowers, Turkey,
28 February and 1 March 1919
Oil on canvas, 36 x 46.1 cm
AWM ART02838

One rainy day, Lambert set himself up with a still life of late winter wildflowers and plants typical of the area. He set the arrangement in a biscuit tin on top of a bed and worked on it while the rain pelted down. The still life may have been a convenient subject to fill in a couple of days, but Lambert was as attentive to the small details in his pictures as to the broader landscape settings. The anemones and euphorbias around the campsite gave character to the landscape, and the artist wanted to make sure that the details in his paintings were authentic. For this reason he intended to do a series of watercolours of the bushes and plants of the area as a way to understand the Gallipoli landscape, but he only completed one of a common arbutus shrub.

While the weather difficulties provided Lambert with interesting aesthetic and technical problems, his initial misgivings about not being on Gallipoli at the right time of year were perhaps well founded. For an artist so convinced of the need for authenticity in his art, working in winter rather than spring was a serious problem. The dull light and extreme weather conditions affected the way he painted. Paint dried slowly in the cold weather and he had to devise a system of storing the half-painted wood panels so that they did not smudge. He could not work in the field on rainy days and the icy conditions numbed his hands. Importantly, the absence of bright sunlight rendered colours into muted tones and eliminated shadows that helped to define landscape features. In contrast to the paintings Lambert made in Palestine, where there was brilliant light and intense colour, his Gallipoli paintings are subdued and he generally used a muted colour palette of mauve-greys, sage greens and ochres.

George LAMBERT (1873–1930)
Study of Arbutus shrub, Turkey, 2 March 1919
Watercolour and pencil on paper, 37.2 x 25.4 cm
AWM ART02860

The details of the landscape fascinated Lambert as much as the battlefield sites. On the day he made this watercolour sketch, he rode over to the Graves Registration Unit camp and in the afternoon 'did a water-colour of the first of a series of bushes and plants which [he thought] should be recorded; a study of arbutus, rather a wax-like leaf with a sort of blossom something like a laurel but with red stalks'.

Hubert WILKINS (1888–1958)
Wild flowers on Gallipoli Peninsula,
14 February–10 March 1919
Paget colour plate
AWM P03631.235

Lambert preferred to paint on days when there was strong sunlight, as it helped to define landscape features, but he was hampered by overcast conditions. On the day he made this painting of the Sphinx from near the landing, the dull light 'left little else than a fine sky-line and flat tone against a grey sky'.

George LAMBERT (1873–1930)
The Sphinx, from Suvla side, grey day, Turkey, 16 February 1919
Oil with pencil on wood panel,
22.9 x 35.7 cm
AWM ART02851

On 27 February the Australian Historical Mission visited Gaba Tepe. In 1915 Turkish guns stationed here had constantly shelled the ANZAC area. Lambert was keen to record the view that the gunners would have had and 'managed to get a small painting of the stuff one sees of Anzac from Gaba Tepe'.

George LAMBERT (1873–1930)
ANZAC, from Gaba Tepe, Turkey, 27 February 1919
Oil on wood panel, 13.7 x 22 cm
AWM ART02825

William SWANSTON (1881–unknown) and
William JAMES (1887–1972)
The Australian Historical Mission at lunch on Hill 60, 22 February 1919
Copy negative
AWM A05258

From left to right: Herbert Buchanan, Zeki Bey, Hubert Wilkins, Charles Bean and George Lambert.

Lambert may not have made exact studies for his planned large paintings, but he painted around the ANZAC area, recording the details and scenes that would later help him to determine colour palettes, approaches and atmosphere. When the Historical Mission visited Gaba Tepe, he sat on a promontory, looked back at the distinctive ridge lines and painted *ANZAC, from Gaba Tepe*. This small and finely crafted work presents an apparently benign and picturesque view, the deep blue of the ocean contrasting with the sandy coloured ridges. However, the site was significant for Bean's assessment of the campaign. The task for the Historical Mission on that day was to examine a Turkish gun position to determine what the Turks had seen as they pounded ANZAC Cove with artillery shells. Lambert's painting provides a visual statement of the day's work. Photographs taken on the same day by Wilkins from a position similar to Lambert's, show the rear of the formidable gun positions and a detailed sketch drawn by Bean of the ANZAC area seen from the crest of Gaba Tepe indicates the importance placed on this visit.[31]

Lambert's Gallipoli field sketches are essentially straightforward records of the scenes and details that he wanted to recall when he came to work on the large commissioned paintings. But as an artist, he also wanted to capture something of his emotional response to the place. If photographers such as Wilkins were to provide the 'documentary truth', then artists were expected to make a personal, and at times emotive, interpretation that would reflect a different sort of truth. Lambert's task was to visit the site and gather material from a range of sources. He would later synthesise these stories and visual cues into symbolic images to convey his version of the 'emotional truth' of key events.

Hubert WILKINS (1888–1958)
A view of a long beach looking from the north of Gaba Tepe, showing the wire and entrenchments between this point and the Australian positions, the outposts of which were at the extreme left, 27 February 1919
Paget colour plate
AWM P03631.230

George LAMBERT (1873–1930)
Burnt gully, Gallipoli, Turkey, 18–21 February 1919
Oil on wood panel, 13.6 x 22.2 cm
AWM ART02824

Hubert WILKINS (1888–1958)
The gun position on Gaba Tepe. The Aegean Sea is in the background, 27 February 1919
Glass half-plate negative
AWM G01956

On 27 February the Historical Mission visited the Gaba Tepe area to investigate where the big Turkish guns 'Beachy Bill' had been hidden during 1915. Wilkins's photograph looking north towards the ANZAC area closely parallels the view painted by Lambert on the same day.

Charles BEAN (1879–1968)
Imbros Island and Samothrace Island at sunset, 14 February–10 March 1919
Nitrate negative
AWM G02087

Many times during the 1915 campaign, Bean and other soldiers had watched the glorious sunsets over the Aegean Sea as they ate their evening meal. In 1915 he had made a small watercolour sketch of the sunset scene; when he returned to the area in 1919 he took photographs of the same view.

Bean took terrain notes as he examined the ground around Leane's Trench, the site of an Australian-led raid on 31 July 1915 to capture a Turkish trench in front of Tasmania Post.

Hubert WILKINS (1888–1958)
Right of Leane's Trench showing the cornfield on top of the hill and Lone Pine in the distant centre, c. 26 February 1919
Glass whole-plate negative
AWM G01943

CHAPTER FIVE

WILKINS AND BEAN
TRACE THE EVIDENCE

THE HISTORICAL MISSION WAS ONE OF THREE AUSTRALIAN groups undertaking systematic surveys of the ANZAC area. Along with the Graves Registration Unit and the AWRS, it engaged in immediate and practical projects that documented the landscape. The diligence and care with which they undertook these tasks demonstrate the importance of the projects and the sense of personal responsibility that the men felt towards the area. The presence of Australians in early 1919 also symbolically reclaimed the battlefields and helped to establish an enduring Australian connection to Gallipoli.

Though the Historical Mission's main aim was to gather evidence related to Australian endeavours of 1915, it was important that the historical and photographic survey of the area that they undertook be completed prior to collecting objects from the field. Charles Bean wanted to see every item *in situ* as he was convinced that the location of objects could reveal much of what had happened: 'what high morning hopes, what grim midday obstacles, and what final tragedy do those cartridge cases or that fragment of uniform tell?'[1] So the Historical Mission group, often with the aid of British or Australian troops stationed nearby, formed search lines and walked the ground. Whether it was kit and cartridge cases, personal relics or bones, they meticulously recorded where any evidence of Australian presence remained and identified those items that seemed most suitable for museum display.

Bean introduced a system of notating the area to help the other groups do their work. Numbered stakes outlined positions held by Australians up to 7 August 1915 in the 'old ANZAC' area and ran alongside another system to identify the ground held after August. The map-makers, Herbert Buchanan and George Rogers, worked in tandem with Bean. Places that had hitherto been roughly located in a general area were plotted and numbered on small field sketches to be transferred later to larger formal maps.[2]

John Balfour, William James and the AWRS team were responsible for documenting the selected relics. Establishing the provenance of an object was critical to its interpretation and future display potential. The exact location where each item was found was logged, items were numbered, and then packed and made ready for transport back to Australia. Cyril Hughes's Graves Registration Unit was notified immediately when bodily remains were discovered so that they might be identified before a decision was made on whether to move them to a consolidated cemetery site. With all this work going on, it was as though the very ground was being archived to form part of the nation's memory.

The Australian men from the Graves Registration Unit probed the ground with steel rifle rods to find burial sites. Once located, these graves were marked with pegs, then surveyed and recorded on charts. Bodies in some outlying areas were moved and consolidated in new cemeteries, headstones were eventually installed and the areas landscaped in preparation for visitors.

Hubert WILKINS (1888–1958)
The graveyard in Brown's Dip, Gallipoli. Lieutenant Cyril Hughes, left, and Sergeant Arthur Woolley marking out graves, February–March 1919
Glass whole-plate negative
AWM G01935

PHOTOGRAPHS AS THE PLAIN, SIMPLE TRUTH

The photographs that Hubert Wilkins made in 1919 were an integral part of the web of evidence that Bean constructed. Bean's training as a lawyer and experience as a journalist had impressed on him the principles of evidentiary proof. During his years as a war correspondent, and after countless interviews with men who had participated in battles, he had developed a thorough method of historical research. All forms of eyewitness report, whether individual testimony, material artefacts or images, were helpful in cross-referencing details and getting as accurate an impression of an event as possible. Wilkins's photographs, when combined with the works of art, notes, maps and objects collected by the AWRS and the Historical Mission, would build a comprehensive representation of Australia's involvement in the Gallipoli campaign.

Wilkins's role as photographer on Gallipoli was similar to that on the Western Front as he followed the battles of 1917–18. Although he had assumed a fair degree of flexibility in his approach, choosing to seek his subjects and follow the fighting as he saw fit, Bean provided almost daily guidance and set an example of personal commitment and tireless energy. In a self-effacing statement made after the war, Wilkins acknowledged this guidance in a letter to Bean: 'We wielders of the mechanism were only agents of your directive influence and were it not for the fact of our great admiration of you and [your] outstanding example, the efforts we put forward would not have been so energetic.'[3] Once on Gallipoli, the familiar working relationship between the two men resumed. They worked together most days, with Bean suggesting many of the subjects for the photographs, but Wilkins assuming responsibility for all technical aspects, the composition and the central focus of the image. As the photographer, Wilkins ultimately applied to each image his own interpretation of the scene.

William JAMES (1887–1972)
Hubert Wilkins, with two Turkish soldiers at Baby 700, c. 17–19 February 1919
Gelatin silver print
AWM P07906.043

Many Turkish soldiers remained on Gallipoli following the Armistice. Some were abandoned by their officers and were forced to salvage fuel and supplies from around the battlefields. Often they had inadequate rations and helped out around the British and Australian camps to obtain food. This photograph also demonstrates the portability of Wilkins's Thornton Picard camera, which he holds in one hand, with his tripod in the other.

For a medium that relied on light to form the images, the weather was a primary factor and work often had to be suspended or cancelled owing to unfavourable conditions. Poor and variable light meant that Wilkins had to overcome technical difficulties to ensure the success of the photographic record. The task of the photographer was a delicate one. Cameras were complicated and required dexterity for loading and unloading the glass-plate negatives. Wilkins had only a limited number of glass plates and the risk of damage was always present. The late winter weather produced days that were generally overcast, with some sleet and snow, and many rainy spells that hindered the group's work.

Wilkins's approach to making the photographic record was relatively straightforward. On a practical level, he decided where to position the camera to best represent the field of action, then determined the framing, exposure, focus and level of detail. He would then load up his glass-plate camera and start taking photographs. He mostly focused on the details of the battlefields but on a few occasions lifted his eyes to take in the wider landscape. Wilkins also carried with him a small number of Paget colour plates, an early form of colour photography that combined a colour screen with a black and white negative. The advantage of this process was that the plates could be used in the same camera and the resultant images could be printed as black and white or colour prints. At the end of each day he would work late into the night in the makeshift darkroom. It was important that the day's plates be developed each evening, to make sure the images were properly exposed and viable. The photographic record was intended to document the marks that remained on the landscape, so it was essential that the images be as accurate as possible. By locating an event or object in an actual landscape, Wilkins's photographs provided supporting evidence for Bean's interpretation of events. Once people could 'see' where something happened, they might more easily understand 'how'.

To achieve images that served this purpose, Wilkins continued to approach his work in the same way he had on the Western Front. His Gallipoli photographs are simply constructed; the compositions are straightforward, with the focus of attention usually in the centre. This all helps to make the photographs easily understandable. It is possible to look at Wilkins's Gallipoli photographs and instantly recognise the place, the detail or the human figure as a literal representation of the subject. This was what Bean wanted for the historical record – images that viewers could trust as reliable documentary records. They were as Bean said, 'a sacred record – standing for future generations to see for ever the plain, simple truth.'[4]

But a photographer contributes much more to every image than just the technical skill required to take the picture. Wilkins brought to his work personal aesthetic judgements and extensive experience in producing complex and evocative images. As a result, they not only serve as records, but also stimulate viewers in other ways. His photographs are full of secondary narratives and meanings, including specific references to the important work of the Historical Mission going on around him.

Hubert WILKINS (1888–1958)
Panorama from the trenches east of Johnston's Gully at the edge of Wire Gully, looking along Marine Trench and the head of Bridges Road, c. 1 March 1919
Glass half-plate negative
AWM G01988P

The Historical Mission investigated the evidence of the Australian party that had raided Gaba Tepe on 4 May 1915; Wilkins took photographs of what the Turks could have seen of the beach where the Australians landed. The barge was washed ashore from the ANZAC area after the raid.

Hubert WILKINS (1888–1958)
The beach from Gaba Tepe, 27 February 1919
Glass half-plate negative
AWM G01967

Hubert WILKINS (1888–1958)
The drain in Shrapnel Gully (made in November 1915 to preserve the roads from winter rains) as it appeared after the Armistice, c. 24 February 1919
Glass half-plate negative
AWM G01950

PRESENCE AND ABSENCE

At a cursory glance, many of Wilkins's photographs appear static and un-peopled, yet they are in fact full of human presence. On the Western Front part of his work had been to make a record of Australian soldiers at rest. He took photographs of exhausted men just returned from battle, billeted in farm buildings or just resting behind the front lines. These were the all-important human participants, the individual characters and the subjects of stories that contributed to the totality of Australia's war effort and experience.

On Gallipoli, Wilkins continued this theme. He frequently included members of the Historical Mission in his photographs, recording moments of intense discussion between Bean and Major Zeki or showing Bean standing over a trench studying his notes. On occasion, George Lambert is shown working on his paintings in the flat fields of Krithia or tucked in among the buildings at Helles. Unlike the photographs taken by Bean or William Swanston and James, Wilkins frequently included his colleagues in the image, highlighting his particular interest in the human figure in the landscape. These photographs also document how the group went about its work.

Wilkins also made more deliberate portraits of the group. On 25 February at Lone Pine he took a photograph to illustrate the closeness of the opposing trenches, which according to Bean were less than 4 metres apart. Wilkins took the photograph as a way of measuring the distance.[5] A second photograph taken on the same day shows five men standing on the narrow bridge of land that separated the trenches. They are acutely conscious of the photographer's attention and all look straight at the camera. The presence of the Historical Mission men in this second image records more than their work on that particular day – the scale of their bodies also emphasises the narrowness of no man's land. Photographs such as this allowed those who had lost loved ones at Lone Pine to see how close the fighting lines were and to vividly picture the terrible struggle in the trenches. And soldiers who had been involved in the brutal hand-to-hand fighting may well have been prompted by Wilkins's photograph to replay the scene over again in their imagination.

Bean and Wilkins were accompanied most days by Buchanan, who was assigned to the Historical Mission as the topographic expert. An ex–railway engineer who had been in charge of mapping for 1st ANZAC Headquarters in France,

Bean and Zeki together walked the ground and conferred on details of troop positions and the events of 1915. Bean wrote that Zeki, a Salonika Turk about 29 years old, was intelligent and brave with a slightly reserved manner. He worked with the Historical Mission for seven days before returning to Constantinople.

Hubert WILKINS (1888–1958)
Looking from Turkish trenches near German Officers' Trench to our trenches, 24 February 1919
Glass whole-plate negative
AWM G01940

Bean found Tommies' Trench almost exactly as it had been in 1915 'with a meaningless collection of holes behind it. Many of these have been filled in since by burying our men in them; others lie in the hedge like rows of tufted scrub.' Lambert is seen in the centre making a sketch and another man, probably Buchanan, is standing in the trench.

Hubert WILKINS (1888–1958)
Panorama from Tommies' Trench showing Achi Baba and the French position on the right, 8 March 1919
Glass half-plate negatives
AWM G02041P

Bean asked Wilkins to photograph the narrow gap between the Australian bombing post (lower left) and that of the Turks (centre right). The Turkish monument in the background marks the spot where the ANZAC advance in August 1915 was stopped by the Turks at Lone Pine.

Hubert WILKINS (1888–1958)
The remains of overhead cover and a mine crater nearby from which they used to throw bombs at Lone Pine, 25 February 1919
Glass whole-plate negative
AWM G01939

Members of the Historical Mission standing on the narrow bridge that separates the trenches. From left to right: Herbert Buchanan, Charles Bean, probably William James, Major Zeki and George Rogers.

Hubert WILKINS (1888–1958)
The distance between opposing trenches at Lone Pine, 25 February 1919
Glass half-plate negative
AWM G01946

Buchanan was there to produce accurate maps of the area. But in 1919 it seems that his duties extended well beyond map-making and he worked closely with Bean as his field assistant. Nearly always neatly turned out in riding breeches and slouch hat, Buchanan, with his tall and lean figure, features frequently in photographs taken by Wilkins. He is shown, for example, at German Officer's Trench examining a concrete-rendered trench (probably on 24 February). The photograph provides evidence of the roughly built trenches and fortifications that modified the landscape for war. It also provides a glimpse of the spaces that the enemy inhabited, yet superimposed over this history is Buchanan's overtly Australian figure. In this way, the photograph reasserts Australian presence in the contested landscape.

Wilkins often took photographs that placed the viewer in the soldier's shoes, showing what the soldier might have seen and thereby assisting viewers to understand their story. Because the images were apparently simple and full of pertinent detail, people could project their own stories onto the scene. And although the viewers were not there in 1915, the visual clues help them to imagine what it may have been like. Importantly, by taking the soldier's perspective, Wilkins's method mirrored Bean's approach to military history, with its concentration on the stories of individuals. This bound into a cohesive narrative the photographic record and the historical record. With each part of the story following the same principles, whether it was in potential museum displays or the official histories that Bean would later write, the stories of Australian troops would be presented in a consistent and interrelated way.

The Historical Mission made its first visit to the area around the Nek on 17 February. There, as we have seen, Hedley Howe recounted the events of the landing and the subsequent rush of men over the ridges. A double-plate panoramic image of the Nek – most likely taken by Wilkins on that day – locates the small field in a broader context, with Battleship Hill and the outline of Baby 700 dominating the skyline. Wilkins positioned his camera close to where Lambert must have been painting, just behind the Australian lines, looking across to the recently erected Turkish memorial that marked the opposing trenches, between 15 and 50 metres away. Even though the panorama format takes in a much larger sweep of the landscape than Lambert's painting, the narrow piece of no man's land between the trenches, which Bean described as being about the size of three tennis courts, dominates the image. The multiple rows of Turkish trenches beyond are clearly visible and indicate the dominance the Turks had over the ground. But, as in so many of Wilkins's images, it is the foreground detail that captures our attention and stimulates our imagination. His photograph makes it possible to see inside the trenches where the Australians had waited before leaping out in that mad dash across the ground. Fragments of clothing are caught on the parapets and, just out of the trench, bleached bones and a skull are physical reminders of the great loss of Turkish and Australian lives.

In photographs of Quinn's Post, the proximity of the fighting lines is again emphasised. The deep gullies and steep ridges that seem to fold into each other are here even more compressed than elsewhere. This is accentuated in the black and white image of the extreme northern end of Quinn's Post,

In mid-1915 Major Zeki had been stationed at German Officers' Trench. He explained to the Historical Mission how the Turks had reinforced damaged walls after the Australians had mined and set off bombs under the trenches.

Hubert WILKINS (1888–1958)
Lieutenant Herbert Buchanan of the Australian Historical Mission standing next to the concrete-faced trench in German Officers' Trench, Gallipoli, c. 24 February 1919
Glass half-plate negative
AWM G01929

When making this panorama, Wilkins set up his camera and tripod close to where Lambert was painting. Wilkins's photograph emphasises the closeness of the opposing trenches and the strategic advantage of the Turkish positions.

Hubert WILKINS (1888–1958)
Looking across the Nek from the Australian trenches to Baby 700, 17 February 1919
Glass half-plate negatives
AWM G02013P

where Turkish trenches adjoined those of the Australians. In 1915 this area changed hands a number of times, eventually becoming a position from which Turkish snipers covered the Australians. The photograph taken by Wilkins from the Turkish perspective, looking back across Australian lines, indicates the clear view a sniper would have had of Australian positions, including the head of Monash Gully, Russell's Top, Pope's Hill and nearby Quinn's Post.

This viewpoint is replicated in an extraordinary image of a British sniper's position on Rhododendron Ridge. Wilkins's camera was placed to record where the sniper would have been hiding. The photograph shows the spiky bushes on either side that constricted the view so as to give only a glimpse of the Turkish-held territory across the gully. A shallow depression in the soil marks where the sniper may have lain in wait, and it is easy to imagine him peering across the narrow gap anxiously watching for any movement. Further evidence of the sniper's trade is clearly visible in the spent .303 cartridge cases scattered around the hollow.

The question of how far Australian troops had pressed forward on the first day was a primary riddle for Bean to solve. The Historical Mission's preliminary surveys had shown that much of the trench system remained intact and could be traced to the further reaches of the advance. Beyond the furthest trench, the remains of Australians scattered through the brush, or hastily buried or burnt bodies, provided the clues. Bean worked over the ground like an archaeologist and Wilkins photographed the objects they found *in situ*. On the hilltop of Baby 700 Wilkins took a photograph of a pierced water bottle resting among bleached bones and fragments of uniform. This was close to the spot that an advanced Australian party had reached on the first day of the campaign. The military objectives of Chunuk Bair and Battleship Hill can be seen on the horizon in the photograph. While Bean reasoned that it was possible for an undamaged water bottle to be picked up and used elsewhere, 'a bottle with holes in it was no use to anyone' and would have been left where it had fallen.[6] When combined with eyewitness testimony, such small objects left on the surface of the landscape enabled Bean to plot out where Australians may have reached and suggested how they were overcome.

Unlike the AWRS photographers, whose work was mainly about gaining a strategic overview of the landscape, Wilkins did not try to explain the grand battle narratives. Instead his photographs document the personal spaces where soldiers lived and died; they represent the often unknowable stories of individuals facing death and trying to survive. Photographs of remnants of kit and clothing tangled in the barbed wire or undergrowth convey narratives that are open-ended, inviting viewers to speculate on how these things came to be there.

Hubert WILKINS (1888–1958)
A sniper's post on our side of Rhododendron Spur, Gallipoli, c. 23 February 1919
Glass half-plate negative
AWM G01911

Empty .303 cartridge cases lie scattered on the ground in the hollow the sniper used for cover.

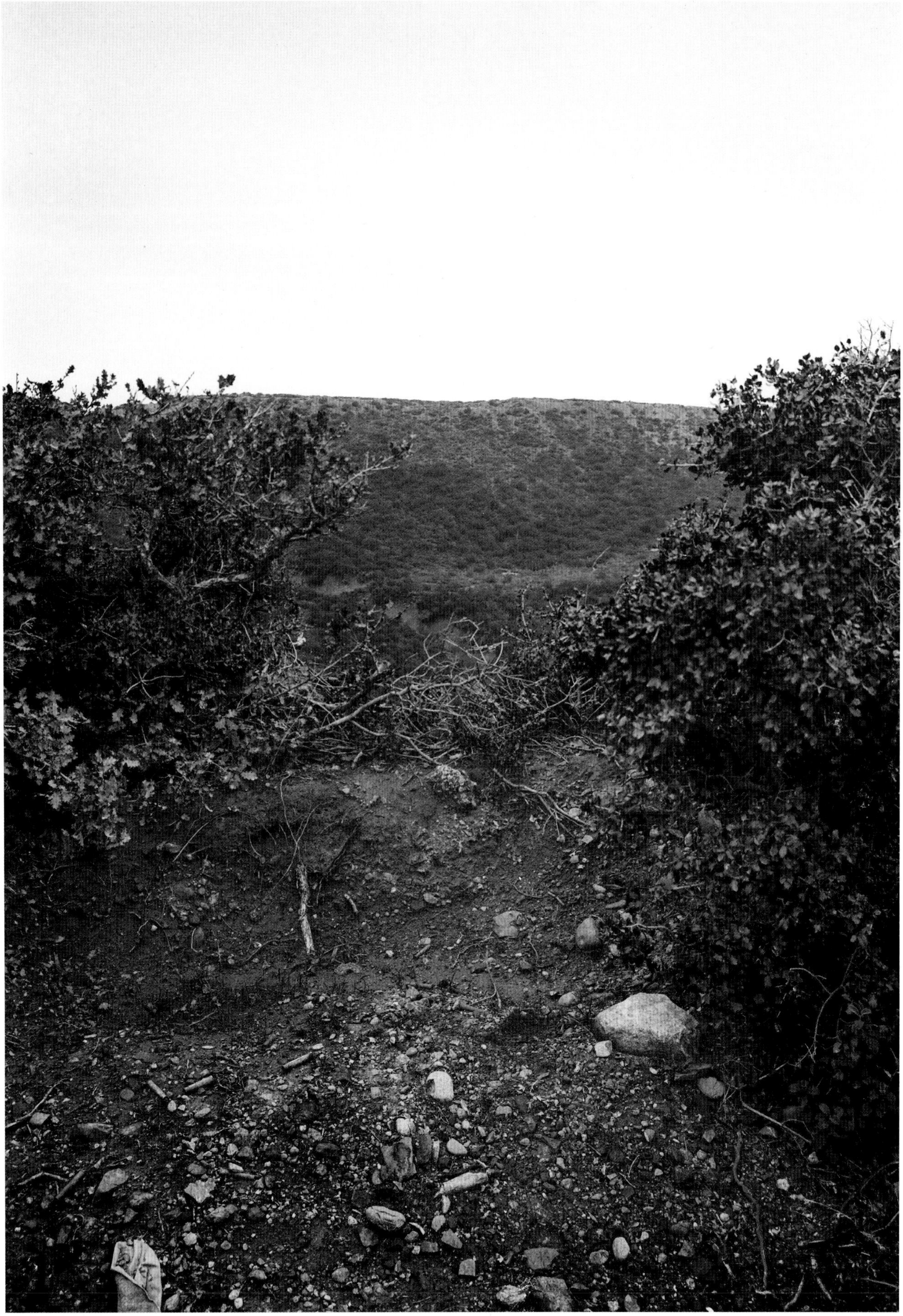

Hubert WILKINS (1888–1958)
Extreme left of Quinn's Post, Gallipoli, showing how our trenches ran into Turkish trenches, c. 24 February 1919
Glass half-plate negative
AWM G01925

Photographs of pieces of bones, kit and clothing provided evidence of the advance parties of Australian and New Zealand troops on the day of the landing in 1915. Bean reasoned that an undamaged water-bottle might be salvaged and carried off by other soldiers but a damaged bottle was of no use and would be left where it fell.

Hubert WILKINS (1888–1958)
The grave of a 1st Battalion soldier on Baby 700, looking towards Battleship Hill,
c. 17–19 February 1919
Glass half-plate negative
AWM G01885

Human bones lying on the surface of the landscape were more tangible markers of trauma and thus became an important motif in Wilkins's photographs of the area. The brief ceasefire on 24 May 1915 had allowed the bodies of many men from both sides to be buried. But those who had been killed in no man's land during the rest of the campaign lay where they had fallen until the Graves Registration Unit started its work in late 1918. When the Historical Mission arrived in February, several hastily filled-in Turkish mass graves had been eroded by recent rains and the bones exposed to the elements and marauding dogs. One of these pits was on the southernmost end of the ANZAC area, near Leane's Trench, where the Turks had repeatedly tried to break the Australian line. The bleached bones could be seen 'a mile away as a broad white streak down the ravine'.[7]

On 22 February Zeki accompanied the Historical Mission on a visit to Hill 60, where combined allied forces had tried to take a small knoll in August 1915. Zeki had commanded Turkish forces here and could testify to the position of the machine-guns that had destroyed the 4th Australian Brigade's attack of 8 August 1915. The Historical Mission found the earth still scorched by a fire that had broken out during the initial shelling. The intensity of the heat had virtually sterilised the soil and little had regrown in the ensuing years. Blackened stumps of trees dominated the scene. As the group surveyed the area, they found ground still littered with relics. A colour patch from the Australian 14th Battalion and a small bible with 'H. Wellington' inscribed on the flyleaf were found there.[8] But the bones of the Australians were everywhere, 'some laying [sic] as far up the ridge as Hill 100 – Australians and Turks together'.[9] Wilkins took a photograph of a small cluster of bones on Hill 100, standing out white on the black landscape.

BEAUTY IN THE BATTLE LANDSCAPE

The mood in the Historical Mission camp was often bleak. The work of the Graves Registration Unit was a constant reminder of the tragic loss of life on Gallipoli. The weather was gloomy and the nickname given to the ANZAC area, 'Spooks Plateau', certainly didn't help to lift their spirits.[10] Lambert wrote that the place was 'damn melancholy' and beautiful as it was, it still depressed him.[11] Bean also noted the sadness that gripped them all, even Wilkins. Of the nearly 200 images Wilkins took on Gallipoli, the majority are close-ups of battlefields, bones and the detritus of war. The sadness of the place is further captured in images he made around Quinn's Post or in Shrapnel and Monash gullies, where the steep folds of the ANZAC area create an almost claustrophobic feeling.

However, Wilkins was not immune to the beauty of Gallipoli, as a small group of photographs, including the Paget colour plates, indicates. On occasion he deliberately sought to create attractive images of the landscape and vistas that were the backdrops to the battlegrounds. Gallipoli's sharp ridge lines and deep valleys, seen against the blue of the Aegean, were ideal for picturesque photographs, and the high points of Chunuk Bair, Rhododendron Ridge and Plugge's Plateau provided vantage points from which to survey the whole of the ANZAC area. These photographs reveal the structure of the distinctive landforms – such as the jagged, crumbling features of the Sari Bair Range –and dramatically convey the rugged terrain that surrounded the ANZACs. Wilkins must have been delighted by the definition he could achieve in his images on days such as these when there was bright light to rake across the surfaces and reveal the landscape formations.

Most of these photographs could still fulfil the required function of constituting evidence. For instance, the image looking across the top of Sniper's Nest towards Suvla Bay shows the structure of Turkish basketwork revetments. Such photographs also helped locate stories about individuals and places in a broader landscape setting. Wilkins found the view from Silt Spur particularly appealing; he took a series of photographs of the expansive panorama of the country and the indented shoreline stretching away towards Gaba Tepe and the southern toe of the peninsula (not visible from the ANZAC area). Another photograph replicates the view, but includes Buchanan looking at a flowering fruit tree. This image captures the lasting beauty of the landscape and the natural cycles of birth, death and regeneration.

Hubert WILKINS (1888–1958)
Bones of Turks and rotting clothing and equipment in front of Leane's Trench, c. 26 February 1919
Glass half-plate negative
AWM G01949

When the Historical Mission visited the Hill 60 area, they found the site littered with bones. Bean recalled that the day of fighting in 1915 had been hot and dry and 'the blackened stumps of the scrub set on fire by our shells still marked the slopes'.

Hubert WILKINS (1888–1958)
The burnt bones on Hill 100 near Hill 60,
c. 22 February 1919
Glass half-plate negative
AWM G01906

Hubert WILKINS (1888–1958)
Looking from a point near the beach between No. 1 and No. 2 Outposts towards Baby 700 on the extreme right, where the bodies of Australians killed in April 1915 were found, c. 22 February 1919
Glass whole-plate negative
AWM G01902

An important task for Wilkins was to record what the Turks could see from their positions. During 1915 a Turkish machine-gun had been positioned on Sniper's Nest and at night fired into the northern side of Ari Burnu promontory about 1,300 metres away. The vessel is the *Milo*, sunk off North Beach on 26 October 1915 to provide a breakwater.

Hubert WILKINS (1888–1958)
Looking from Sniper's Nest to Ari Burnu, c. 2 March 1919
Glass whole-plate negative
AWM G02019

When Wilkins slipped the Paget colour plates into his camera, he was more deliberately seeking to make beautiful images, but the capacity for the photograph to be used as evidence was diminished. In the early twentieth century, the stark clarity of black and white was the preferred medium for documentary photographs. Colour photographic technology was in its infancy, and the Paget system only approximated the true colours of nature. The colour photographs made on Gallipoli in 1919 more closely resemble colour sketches that might have been made by an artist. Tones are generally muted yet blues are accentuated, which brings out the detail in the sky. A photograph taken from Quinn's Post looking to Hill 971 around 25 February illustrates this point. The foreground detail is not as sharp as in the black and white version and some of the tonal contrasts are subdued. Our eyes are drawn to the massive cumulus clouds moving across the sky. This image, with its emphasis on the cloudscape, is reminiscent of a John Constable sketch and is in stark contrast to several black and white photographs that Wilkins took around the same spot.

THE RECORD AND COMMEMORATIVE PHOTOGRAPH

As we know, the photographic record that Wilkins made on Gallipoli in 1919 was intended to provide visual evidence of the military events of 1915 and to support the historical narrative being developed. Even though nature had started to reclaim the scarred earth and the trenches were starting to erode, the signs of battle were everywhere. Wilkins's Gallipoli photographs stand as an eyewitness account of the traces of battle that remained on the landscape three and a half years later. Importantly, they locate individual objects and stories and the work of the Historical Mission within a specific landscape.

In addition, these photographs were part of the primary resources being gathered for the planned official histories and would also be used in a range of related publications produced by the Australian government through the interwar years. As they would be available for purchase by the Australian public, they needed to be truthful. Their value as a 'scrupulously accurate photographic record' was highlighted by Bean in the first volume of the official history, published in 1921.[12]

By 1919 it was known that Wilkins's photographs were not going into a military museum but to a new form of institution that had the dual functions of a museum and a national war memorial. In fact, all the material collected by the Historical Mission was intended to serve as a memorial to Australia's involvement in the Great War. Bean believed that the work of the official Australian photographers should have 'proper regard for the great sacrifices, and the sacred memory of the great men to whose bravery these pictures should be a monument'.[13] The war and Australia's contribution to it had been an historic event that demanded to be understood, but it was equally important that the sacrifices made by individuals and the enormous emotional impact on the nation be remembered and acknowledged. When confronted with the scale of the loss of life that had occurred on the Western Front and the tragic events on Gallipoli, Wilkins used his skill as a photographer to make images that would help to ensure the Australian people never forgot these events.

Bean described Sniper's Nest as 'this solitary trench along the razor edge amid a billowing sea of wild, steep spurs. It was revetted with basketwork, the circular tops of the baskets holding up the parapet, and the space between each basket and the next serving as a loophole.' Suvla Bay and Salt Lake can be seen in the distance.

Hubert WILKINS (1888–1958)
The remains of revetments on the parapet of the trench leading to Sniper's Nest, c. 2 March 1919
Glass whole-plate negative
AWM G02020

At Silt Spur the Australians had constructed a network of tunnels running just below the surface. At short distances a series of concealed openings provided sentries with look-out posts. For Wilkins, the mounds of deposited soil from the tunnels provided a dramatic foreground against the distant headlands of the peninsula and the Aegean Sea.

Hubert WILKINS (1888–1958)
The silt on Silt Spur, c. 26 February 1919
Glass half-plate negative
AWM G01954

When the Historical Mission visited Silt Spur, they found a young fruit tree growing in the heaps of sand generated by Australian tunnelling activities. The flowering fruit tree had possibly grown from a discarded stone from jam sent to Australians; it provided the opportunity for a picturesque photograph. Hearing the story, some veterans were prompted to remark, 'There were no stones in *that* jam.'

Hubert WILKINS (1888–1958)
Lieutenant Buchanan standing beside a flowering fruit tree on Silt Spur, c. 26 February 1919
Glass half-plate negative
AWM G01951

Hubert WILKINS (1888–1958)
Looking up the Valley of Despair to Wanliss, Surprise and Cooee Gullies, c. 27 February 1919
Glass half-plate negative
AWM G01974

Hubert WILKINS (1888–1958)
Ravine Gully and a cemetery at the old British front lines, 8 March 1919
Glass half-plate negative
AWM G02055A

Hubert WILKINS (1888–1958)
A view showing the road reconstructed by the Turks after the evacuation, leading over Quinn's Post and onto Hill 971, c. 25 February 1919
Paget colour plate
AWM P03631.231

The three ships in the harbour – *River Clyde* (left), a French steamer (centre) and the hulk of the French Navy battleship *Masséna* (right) – were run ashore to form a breakwater and provide some protection for British and allied landing forces in 1915.

Hubert WILKINS (1888–1958)
Helles Beach showing the fort on the extreme left from which the machine guns were fired by the Turks during the landing, c. 8–9 March 1919
Glass whole-plate negative
AWM G02035

The flat country stretching towards the low hill of Achi Baba was described by Bean as covered by 'low, rounded, grey-blue and green shrubs like saltbush, with the dry light pinky earth between them'. It provided little shelter for soldiers. George Lambert can be seen making his sketch of *Achi Baba from Tommies' Trench*.

Hubert WILKINS (1888–1958)
The type of country advanced over by the 2nd Australian Infantry Brigade during the attack on Krithia on 8 May 1915, 8 March 1919
Glass half-plate negative
AWM G02060

William SWANSTON (1881–unknown) and
William JAMES (1887–1972)
Razor Back leading from Table Top to Old No. 3 Outpost.
Chailak Dere on right; No. 2 Outpost on left near sea.
Taken from Table Top looking west, c. 22 March 1919
Glass 10" x 12" negative
AWM G01829

CHAPTER SIX

BATTLEFIELD LANDSCAPES
THE WORK OF JAMES AND SWANSTON

IN ONE OF HIS FIRST PRESS REPORTS FROM GALLIPOLI IN 1915, Charles Bean likened the landscape to the Hawkesbury River country in New South Wales or the gullies around Sydney.[1] He described how the long ridges sent out fingers of gravelly yellow earth which were separated by steep valleys. Covering much of it was 'a low scrub very similar to a dwarfed gum tree'.[2] The comparison to familiar Australian landscapes helped readers to imagine the sort of place where Australian soldiers were fighting. In subsequent press reports, Bean continued to include evocative descriptions of the ANZAC area, providing his readers with 'word pictures' of the overall setting for the ANZAC campaign.

Although elements of the Gallipoli landscape may have seemed familiar to the troops, it was difficult country in which to conduct a military operation. The undergrowth was prickly and the ground loose and crumbling, so that men climbing the steep slopes had to hold onto whatever they could to pull themselves forward. Those landing in the confusion of battle must have found it a disorienting landscape. Standing in the ravines it would have been hard to see where to go next or to understand the relationship of one place to another. From the ridge lines, the ocean and principal landform features provided reference points, but it was terrain that rarely allowed a comprehensive view of the battlefields. Coupled with the inadequate maps available at the time of the landing, this contributed to the disarray of the fighting during those first few days of the campaign.

The photographic record compiled by Hubert Wilkins of places of specific interest for Australians alluded to the lingering and shadowy human marks left on the Gallipoli landscape. The scale of individual sites such as a sniper's post or the narrow gap between opposing trenches was often picked out in some way within the photograph. Wilkins's photographs hinted at the stories of individuals. Yet Bean recognised that for Australians to understand the military campaign they would also need an appreciation of the relationship between these individual sites and the totality of the battlefield.

Although William James and William Swanston, the photographers from the AWRS office in Cairo, were not officially part of the Historical Mission, their photographic project proved entirely complementary to that of Wilkins. James and Swanston worked on the peninsula for three months and they took about 150 photographs with a large format 10 x 12-inch camera. Generally, these were expansive views that provided detailed topographic information about the battlefield landscapes that would help people gain a more comprehensive understanding of the campaign. Combined with about 200 plates that Wilkins exposed, the future National Collection was set up with a full photographic survey of Gallipoli as it was at the end of the First World War.

WAR RECORDS IN CAIRO

James and Swanston were on Gallipoli under the instructions of the AWRS, Cairo. This subsection of the AWRS was established in late 1917 to document the activities of the Light Horse units that had been fighting in Egypt, Sinai, and Palestine since 1916. Henry Gullett, who had worked as a war correspondent alongside Bean on the Western Front, arrived in October 1917 to take charge of the subsection and immediately set about collecting documents, diaries, photographs and relics associated with the desert campaign. His task was made difficult by the resentment he found among the Australian troops, who believed that their efforts in battles such as Romani, Magdhaba and Beersheba were being overlooked in favour of events on the Western Front. Gullett commented to Treloar that 'the Light Horse has reason to be sore about its treatment. No publicity, no trophies, no mention in official communiqués, and a most inadequate representation upon staffs, etc'.[3] He campaigned vigorously to turn this perception around, and the section soon became very active and productive.

From the time of his appointment, Gullett managed the AWRS in Cairo and continued his role as official war correspondent with the Light Horse. Additionally, he was nominated as the official historian of the AIF in Sinai and Palestine. In mid-1918, when it seemed the war in the Middle East was drawing to a close, he formally withdrew from records work so that he could begin gathering material for the official history. But Gullett's roles as correspondent and historian were so closely connected with those of record-keeping and collecting that his active involvement in all aspects of the AWRS continued even after Captain Hector Dinning was appointed as the senior officer.[4] Dinning and Gullett made a good team and worked closely on collecting records and appropriate material for the museum. Throughout the war the program of the subsection continued to be monitored by John Treloar, head of the AWRS in London.

In late 1917 Gullett wrote to Bean suggesting that it was an opportune time to send artists and photographers to Palestine, so in early 1918 the AWRS, Cairo, hosted the official war photographer Frank Hurley.[5] Hurley recorded the activities of the Light Horse units in photographs and film. Shortly afterwards, George Lambert spent six months travelling with the Australian units and visiting battlefield sites such as Romani, Magdhaba and Beersheba. Lambert found travelling with the Light Horse exhilarating. He loved the 'blazing days but cool nights clear lit by either moon or stars' and being surrounded by 'sunbronzed men and beautiful horses'.[6] Successful visits such as these helped to redress the feeling of resentment felt by many of the Australian troops and stressed the importance of the work of the AWRS in the Middle East.

Photography was a large part of the work of the Cairo unit. Unlike the Western Front, here there was no restriction on the use of private cameras and Gullett noted that at any time there were between 500 and 1,000 private cameras with the Australians in the field.[7] This made for an extensive pictorial record, and wherever possible the AWRS copied the photographs and negatives of soldiers for the growing collection. The AWRS had its own small and well-equipped photographic unit that included dark room staff and official photographers such as Hurley, James Campbell and Oswald Coulson, who also documented the work of the Australian units.[8] Reliable supplies of chemicals for processing photographs were difficult to obtain but nearby hospital bases and the RAF photographic division were very helpful.[9] The RAF had a close relationship with the AFC and the photographic section of No. 1 Squadron that had been set up in 1916 to take aerial reconnaissance photographs. When the Cairo subsection was set up, several of its staff and photographers came from the AFC.

Photographer unknown
A photographer trimming prints in the photographic dark room at the Australian War Records Section, Cairo, 1919
Glass whole-plate negative
AWM B01390

From left to right: Lieutenant Murray, survey officer; Lieutenant D.M. O'Connor, photographic officer; Lieutenant Hector Dinning, commanding officer; Lieutenant William James, trophies officer.

Photographer unknown
Officers of the Australian War Records Section, Cairo, 1918
Glass 10" x 12" negative
AWM B01372A

BATTLEFIELD LANDSCAPES: THE WORK OF JAMES AND SWANSTON

From left to right: Trooper William Spruce with a transport mule, unidentified (behind) and Sergeant William Swanston standing with their 10-inch by 12-inch field camera mounted on a tripod.

Photographer unknown
Members of the AWRS photographic team at Gallipoli during early 1919, January–April 1919
Copy negative
AWM P06960.001

THE ANZAC AREA SHOWING MAIN POSITIONS OF 1915

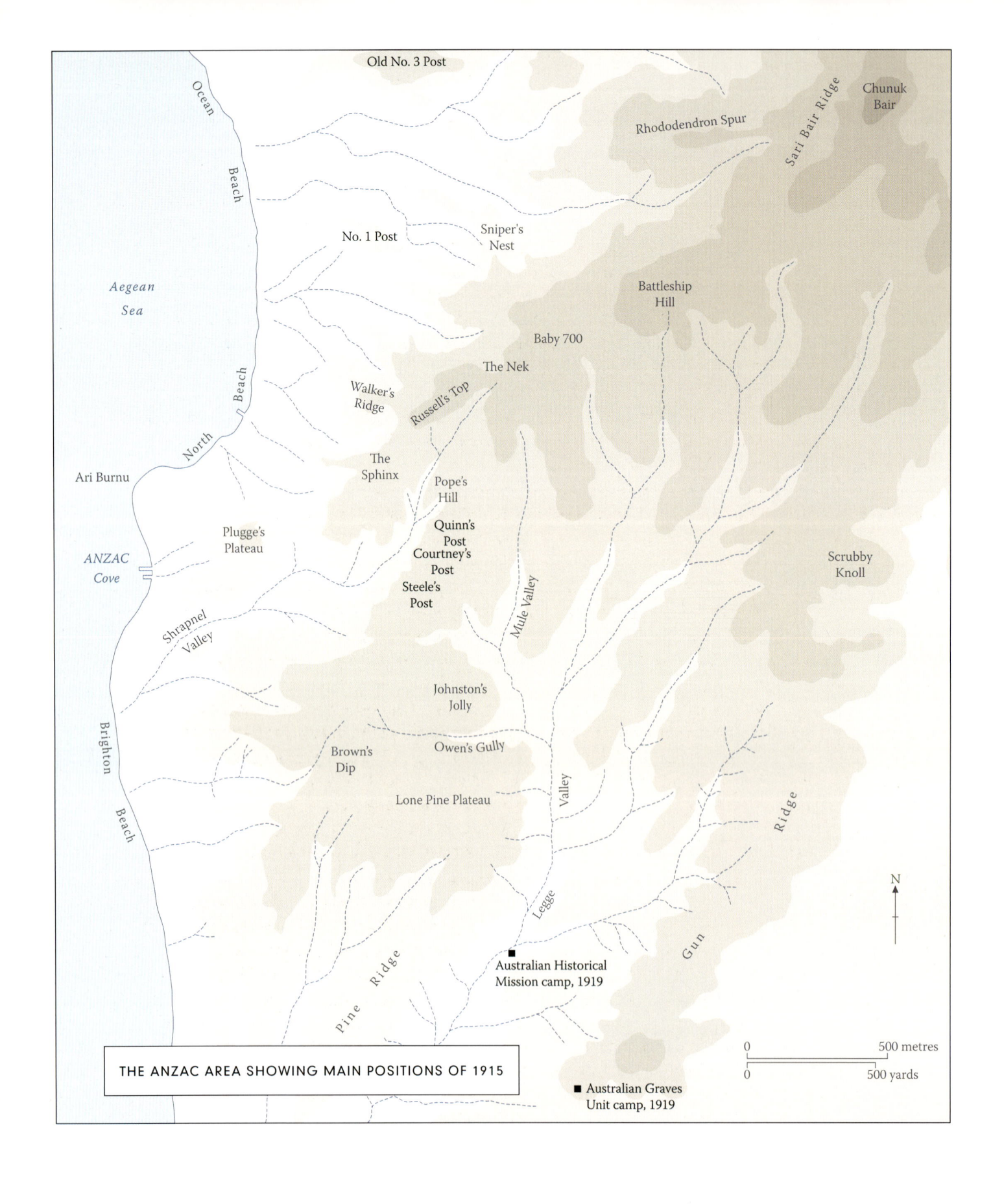

DOUBLE MISSIONS

Bean had set the official wheels in motion in late November 1918 to get permission to lead his expedition to Gallipoli. By the time all members of the party were confirmed and final approvals received, it was January 1919 and their departure for Gallipoli was imminent.[10] In the meantime, the AWRS, Cairo, had already landed a party on the peninsula to collect relics and make a photographic survey of the ANZAC positions. This had been prompted by a cable Treloar sent from London to Cairo on 25 November 1918 advising Gullett of Bean's plans to visit Gallipoli.[11] The very next day, on 26 November, Gullett appointed William James to take charge of an independent group that would travel to Gallipoli as soon as possible to take photographs and collect relics on behalf of the AWRS, Cairo.[12] James was a career soldier and an excellent choice. His service with the 1st Light Horse on Gallipoli from 12 May right through until December 1915 meant that he had good knowledge of the ANZAC area. In addition, he had already worked for the AWRS gathering material from the desert battlefields, so he understood the aims and principles governing the unit's work.

James received his detailed instructions from Dinning. Also a veteran of the Gallipoli campaign, Dinning was a keen advocate for the museum collection. He believed that these objects would be a lasting testament to the work of the specific fighting units and that museum visitors who did not read the written official histories would find the stories around such objects informative and absorbing. Dinning strongly believed that all battlefield relics should have detailed information about their history and connections to Australians as this gave the object its own 'personality'.[13] For him, provenance was everything.

James was directed to collect any material pertinent to the Australian occupation as well as 'enemy souvenirs' that might help tell a more complete story to museum visitors.[14] James was asked to work especially hard to secure any relics 'from areas from which Australians on Anzac were harassed by the Turks – notably for instance, the olive grove which concealed the Beachy Bill enfilading battery, the Anafarta guns positions and the forts of the Narrows from whose gun fire we used to suffer'.[15] It is no coincidence that these were exactly the areas that Bean intended to research when he visited Gallipoli, as they were keys to the unresolved puzzles of the campaign.

As well as collecting relics, James was also asked to supervise a photographic survey of the 'old ANZAC' area. There were two main reasons for this. First, it was intended that James would make a visual catalogue of the Gallipoli landscape. With the war over and the peninsula under allied control, this was the first opportunity since the 1915 evacuation to return and make a systematic photographic record of the ANZAC positions. These images would provide detailed topographic information and form part of the official collection for the museum.

Second, a project had already commenced to make large relief models of battlefields that would illustrate events important to the Australian experience of war. There was a long European tradition of making scaled model replicas of battlefields. Bean had been particularly impressed by the one of the battle of Waterloo that he had seen on display in London, and the Turks had already constructed a relief model of the battle for Gallipoli that was on display in a museum in Constantinople.[16] The photographs James brought back would help the modellers construct accurate sculptural reliefs of the ANZAC area.

James was not a professional photographer but he was familiar with the process and always carried a small pocket camera wherever he went. Immediately after receiving his instructions to go to Gallipoli, he met with officers from RAF headquarters to borrow a camera that could take the detailed images necessary for the official record and for the model-makers. In response, the RAF said the Australians could have virtually anything they wanted and provided the best camera they had, as well as a supply of glass-plate negatives.[17] The RAF camera was said to be the only one of its kind in Egypt and used 10 x 12-inch glass-plate negatives. The definition and clarity available from such a camera was truly remarkable and it would fulfil all the AWRS requirements.

William SWANSTON (1881–unknown) and
William JAMES (1887–1972)
The starting point of the main attack on 6 August 1915 of the Auckland Mounted Rifles. The regiment had a roll call near the (centre) tree on 9 August and only mustered 60 men. This position is in the hollow in front of No. 2 Outpost looking east, c. 14 February 1919
Glass 10" x 12" negative
AWM G01817

William SWANSTON (1881–unknown) and
William JAMES (1887–1972)
Dardanelles with Chanak on the far side from Battleship Hill looking south west. This was taken from the furthest point reached by Australian Infantry on 25 April, 1915,
c. 19 February 1919
Glass 10" x 12" negative
AWM G01827

The RAF also generously made available William Swanston, a highly skilled and experienced photographer.[18] Swanston came from a photographic family: his father started up a professional photographic studio in Edinburgh around 1896 and was eventually joined by two of his sons, who took on the family trade. Prior to his enlistment in the RFC in 1916 Swanston had been a camera repairer and marine photographer working in Aberdeen and Edinburgh. At the time of his secondment to the AWRS, he was attached to No. 14 Squadron in Egypt; listed as a Sergeant Mechanic, he most likely worked with the aerial photography unit. There was also an Australian connection as his mother had been born in Tasmania around 1857.[19]

Swanston possessed excellent photographic skills for the job. Although he had not served on Gallipoli, he was said to have a 'good theoretical knowledge of its geography and of the important Australian positions'.[20] He was given detailed instructions to take notes on the topography and colouration of the landscape for later use by the battlefield modellers.[21]

What evidence exists, such as James's war diary, tells us little about how the partnership between James and Swanston worked. James's experience of Gallipoli gave him an insight into the most important sites to be photographed and he provided direction about the subjects to be recorded. And given his background as a professional photographer, Swanston was most likely the man behind the camera, the one making the technical and aesthetic decisions.

The arrangements for their travel to Gallipoli were made quickly. By 12 December 1918 they had packed up the precious glass-plate negatives and the RAF camera and were waiting in Kantara for the ship that would take them across to Salonica and then to Gallipoli. On 30 December they arrived at the peninsula and set up a temporary camp near the 7th Light Horse work party at the old Turkish Hospital at Cham Burnu. The 7th had been advised of their arrival and provided them with horses, wagons and men to help with the relics work. Just a few days later, James and his party moved their camp to the old ANZAC area, close to the camp of Cyril Hughes and the Graves Registration Unit; it was here that the Historical Mission would also make their base a few weeks later.

William JAMES (1887–1972)
William Swanston at Chanak, February 1919
Gelatin silver print
AWM P07906.038

William SWANSTON (1881–unknown) and
William JAMES (1887–1972)
View looking south down Pope's Reserve Gully (Hill) from the Turkish trenches near Russell's Top, c. 13 January 1919
Glass 10" x 12" negative
AWM G01775

Blizzards in early January made it difficult for Swanston and James to work in the field. In this photograph the pock-marked landscape of trenches and tunnel entrances is highlighted by snow drifts.

William SWANSTON (1881–unknown) and
William JAMES (1887–1972)
Owen's Gully taken from Turkish trenches in front of Lone Pine looking west, c. 12 January 1919
Glass 10" x 12" negative
AWM G01795

SEEING THE BATTLEFIELDS

Like most Gallipoli veterans, James found his return to the battlefields a highly emotional experience. One of the first things he did was to visit the places that were important to him, in particular, Lone Pine and Pope's Hill, where he had been during some of the worst fighting on 7 August 1915.[22] Ordered to attack, the 1st Light Horse had lost many men that day when they tried to take strongly entrenched Turkish positions at Bloody Angle and Dead Man's Ridge. On his return, James discovered that all the posts 'were strewn with the bones of Australia's best'.[23] It was almost impossible to describe 'the tangle of trenches' around Lone Pine and Quinn's Post: 'To stand on the Turkish Trenches at the head of Shrapnel Gully, between Popes Hill and Quinns Post and look down the Gully one really wonders how it was possible to live in the Gully'.[24] On 2 January 1919 he accompanied Hughes to Pope's Hill and found the identity disc of a good friend 'who was killed along with a great many more of my Regt (1st) in an attack on the Turkish Trenches in front of Popes Hill on the 7th August 1915'.[25]

This powerful reminder of loss only served to increase James's commitment to search for what he called 'stuff' that could help explain what had happened. The weather during early January was generally too poor for photography, so with some men from the 7th Light Horse he started to collect relics. Although the Turks had salvaged material from much of the area, within two weeks he had retrieved five watercarts and eight gun limbers as well as many smaller items. He was assiduous in noting the location where an item had been found. The smaller objects were packed in boxes and then moved, along with the larger pieces, down to a keeping area near the Kilid Bahr jetty. While collecting objects, James picked out the places to which he and Swanston would later return to photograph.

When the weather lifted, they started on their photographic survey. The equipment was heavy and awkward and had to be carried on a mule that also transported the delicate glass plates. William Spruce (who also assisted Lambert) was often assigned to help manage the mule and the gear. After reaching the intended site, it took some time to set up the camera on a tripod and work out the focus and exposure times for the conditions. A typical exposure for this camera would have been about 1/30th or 1/60th of a second, and days that had even light and relatively little wind were the best for getting clear images. On their first day, 12 January 1919, they took 11 photographs around Lone Pine and Holly Spur. The next day they worked around Courtney's Post, Quinn's Post, Pope's Hill and Shrapnel Valley, exposing 12 plates. James noted that, with the time taken to travel between the sites and then set up the camera, this was about the maximum they could expose each day. This was in stark contrast to the work of Wilkins, who with his smaller camera could, as James put it, 'let fly all over the place'.[26]

When Wilkins photographed the battlefields he created intense and personal images, using close-ups to evoke the presence of soldiers in particular spaces. This was a more modern approach as it involved the viewer in an emotional experience. James and Swanston's images are very different, and their slightly detached viewpoint is part of an older nineteenth-century picturesque landscape tradition. This is not surprising. Swanston had grown up in Scotland, where the dramatic landscape and coastal regions had fostered the development of a thriving scenic photographic industry in the late nineteenth and early twentieth centuries. As professional photographers, members of the Swanston family probably contributed images to guidebooks, pictorial publications and postcards that were aimed at the well-established tourist market.

William SWANSTON (1881–unknown) and
William JAMES (1887–1972)
Chatham's Post taken from the ridge above Shell Green looking south. The trenches running away to the right were held by Turkish soldiers, c. 21 March 1919
Glass 10" x 12" negative
AWM G01761

William SWANSTON (1881–unknown) and
William JAMES (1887–1972)
Shell Green taken from the side of the ridge near Casualty Corner looking south, c. 21 March 1919
Glass 10" x 12" negative
AWM G01758

'THIS IMPOSSIBLE PLACE'

Many of the photographs James and Swanston made present panoramic overviews of the battlefield that help viewers locate individual events within the larger scene. To understand the way the campaign unfolded after the initial landing, it is essential to know the scale of the landscape and the relationship between specific places. This is evident in photographs such as the one taken from Shell Green looking towards Casualty Corner, which shows the Australian positions along the hill at the back and Clarke's Gully in the foreground. Trenches and tracks cut through the terrain, a small pine tree on the right marks the cemetery, and a Turkish observation post constructed after the evacuation is easily picked out.

A view across Lone Pine looking south from the Turkish trenches shows what James described as 'this impossible place'.[27] Honeycombed with trenches, craters and tunnels, it looks like a lunar landscape. He remarked that 'the whole place is right away from all the rules of war' and that it seemed unbelievable that men had lived in these conditions.[28] It is clear from letters James sent to Cairo that his memories of 1915 were informing the photographic survey. Images such as the view of the back of Quinn's Post trace his connections to the place he called the 'very worst post on the line' and now just a 'mass of torn and broken trenches'.[29] This was where his brigade had had its headquarters, and the photograph reveals the convoluted landforms and the dugouts scraped into the earth by the 1st Light Horse Brigade. The proximity of the Turkish-held positions along the upper ridge is evident.

Wilkins had used five human figures in his photograph to indicate how close the opposing trenches were at Lone Pine, but the pictures taken by the AWRS give a more detached view. Rarely is a human figure inserted into the scene. Instead, they focus on the middle distance and do not include any of that intensely evocative foreground detail – fragments of clothing or bleached bones, for example – that can be found throughout Wilkins's Gallipoli photographs. The latter encourage the viewer to speculate on human narratives, whereas the photographs of James and Swanston provide empirical evidence of a place long since vacated by men.

James was keen to solve some of his own riddles. He made a particular point of taking photographs from Turkish positions overlooking Australian trenches or looking towards the country that lay inland from the places he was familiar with.[30] Some of these provide evidence of what lay beyond the Turkish lines – in many places the land opened out onto gently sloping plains. This was the country that had remained unattainable in 1915. James also tried hard to solve the vexing question of the location of the 'Beachy Bill' batteries that had fired onto the ANZAC area. He spent several days working with officers from the 7th Light Horse to establish their positions. His conclusion was that they were probably hidden in Azak Dere just south of Gaba Tepe and he had Swanston take a couple of photographs of the abandoned gun pits.[31]

Since 1915, the trenches around Lone Pine had been eroded a little by rain but were generally unchanged in early 1919. James noted that the barbed wire fence in the foreground had been recently erected by the Turks.

William SWANSTON (1881–unknown) and
William JAMES (1887–1972)
Lone Pine, looking south from the Turkish trenches on Johnston's Jolly (400 yards away), c. 25–28 January 1919
Glass 10" x 12" negative
AWM G01753

This is where James had served in 1915.
He described it as 'this impossible place'.

William SWANSTON (1881–unknown) and
William JAMES (1887–1972)
The back of Quinn's Post taken from near the 1st Light Horse Brigade Headquarters looking east. Part of Dead Man's Ridge is seen in the top left corner. Turkish troops held all the skyline from the left almost to the centre,
c. 13 January 1919
Glass 10" x 12" negative
AWM G01768

William SWANSTON (1881–unknown) and
William JAMES (1887–1972)
The Dardanelles from Hill 971 looking south-west. Chanak on far side of Narrows in right centre, c. 19 March 1919
Glass 10" x 12" negative
AWM G01825

William SWANSTON (1881–unknown) and
William JAMES (1887–1972)
View of the country behind Lone Pine,
c. 12 January 1919
Glass 10" x 12" negative
AWM G01865

William SWANSTON (1881–unknown) and
William JAMES (1887–1972)
View in Asmak Dere just south of Gaba Tepe. The hollows along the left side of Dere have been used as gun positions facing ANZAC. This is probably where Beachy Bill was hidden. Looking east, c. 12 March 1919
Glass 10" x 12" negative
AWM G01842

While these photographs placed Australian activity within the context of the broader landscape, James and Swanston had also been instructed to gather reference material that could later be used by the topographical modellers to make accurate scale reproductions of the landscape. These models were intended for display in the planned museum. The RAF camera's 10 x 12-inch negatives had the capacity to provide the extraordinary detail needed and panoramic images gave the best information for the modellers to work from. Travelling to a position north of ANZAC Cove, they took a photograph of the landscape between the rising hill of Baby 700, Number One Outpost and Walker's Ridge. The exceptional clarity of this image reveals the astonishing Gallipoli topography, its ridge lines thrown into sharp relief by strong sunlight.

On another day, they set up the camera on the Table Top plateau and took a photograph looking along the sharp line of the Razor Back that led to the old Number Three Outpost. Here is revealed something of the impossibility of the Gallipoli landscape. The slopes of the ridges that face south-west and which generally confronted the ANZAC troops when they landed are deeply eroded by the weather and contain many sheer, gravelly cliff-faces. Those ridges that face north-east slope more gently and are covered with low scrubby vegetation. The AWRS photograph shows each ridge line and track picked out in the bright sunshine and provides an almost scientific view of the landscape.

These images are mostly about fact, not emotion, and the exactitude with which the landscape was recorded was exactly what the model-makers required. This is apparent in a photograph taken from Russell's Top. The beach curves away northwards to Suvla and the salt lake. In the middle distance, a small ship that was beached during the severe early winter storm of 17 November 1915 lies just offshore. Within this one image, the structure of the landscape north of ANZAC Cove is recorded to give the modellers something to work from: the eroded range with its ridges reaching down towards the ocean, the flat hinterland, and the shape of Suvla in the far distance are all laid out in exquisite detail.

Swanston's experience and interest in marine subjects is evident in several photographs of the wrecked hulks around the Gallipoli coastline. The steamer *Milo* was grounded as part of a breakwater in late October 1915, and Swanston uses a calm and formal composition to show the vessel resting in a bay. The image of a Turkish ship off Kilia Liman near Maidos with the strong geometry of the wooden piers in the foreground is also a highly structured image. These photographs of partly submerged ships perhaps indicate the style of photographs that Swanston took before the war, when he was a marine photographer. Qualities of stillness and formality, clarity and precision, are evident in many of the AWRS photographs of Gallipoli.

Like Bean, James and Swanston were accumulating visual evidence and the style of their photographs reflects this approach. The impression one gets from looking through the photographs they created is of a vast and beautiful landscape, silent and motionless. Together with the relics work undertaken by James and his assistants, the AWRS team effectively catalogued the landscape, ordering and scaling its particularities and retrieving physical evidence of Australian experiences on Gallipoli. Although the AWRS party was independent of the Historical Mission and had different objectives, the groups collaborated, sharing resources and constantly exchanging information. While the history and management of each group may have differed, the overall aims – to provide a comprehensive visual record of the Gallipoli landscape and to trace Australian stories that might still be revealed – were common to both.

William SWANSTON (1881–unknown) and
William JAMES (1887–1972)
View of No. 1 Outpost in the centre, Walker's Ridge to the right and Baby 700 to the left. Russell's Top is on the skyline just to the right of No. 1 Outpost with the monument just visible, c. 14 February 1919
Glass 10" x 12" negative
AWM G01814

William SWANSTON (1881–unknown) and
William JAMES (1887–1972)
North point of ANZAC Cove with slope leading to Plugge's Plateau on the left. The south side of the Sphinx is on the right. Taken from Walker's Ridge at the back of the Sphinx looking west, January–April 1919
Glass 10" x 12" negative
AWM G01835

Swanston and James's photographs were taken to help modellers make three-dimensional representations of the Gallipoli landscape. In this image their camera recorded the northern part of the ANZAC area with Suvla Bay and Salt Lake in the distance.

William SWANSTON (1881–unknown) and
William JAMES (1887–1972)
View looking north from the cliff west of Russell's Top, showing No. 1 Outpost in centre right with No. 2 Outpost above it, c. 14 February 1919
Glass 10" x 12" negative
AWM G01776

Swanston's earlier training as a marine photographer is evident in this image of the steamer *Milo*, sunk to make a breakwater on 26 October 1915. A barbed wire fence erected by Turkish soldiers after the evacuation can be seen in the foreground along the edge of the beach.

William SWANSTON (1881–unknown) and
William JAMES (1887–1972)
View of Ocean Beach from the ridge near Ari Burnu looking north-east, c. 22 March 1919
Glass 10" x 12" negative
AWM G01779

William SWANSTON (1881–unknown) and
William JAMES (1887–1972)
View from Kilia Liman on the Narrows near Maidos looking south-west showing the Gap and Hill 971 in the background, c. 28 March 1919
Glass 10" x 12" negative
AWM G01781

THE HISTORICAL MISSION MOVES TO HELLES

By about 4 March the Historical Mission's work around the ANZAC area had nearly been completed. They had traversed the key battle sites several times, made detailed records, and collected and tagged potential museum objects. Bean was pleased with what had been achieved, commenting that 'pretty well every question which we came here to solve has been settled'.[32] But the extended stay in the cold weather and the physicality of the work had affected several members of the group, and there was an air of melancholy in the camp. To make matters worse, Wilkins was unwell, the result of overwork and late-night stints developing negatives in the unventilated makeshift darkroom. Nonetheless, he took a few last photographs and Swanston helped out by making prints from Wilkins's exposed plates.

Lambert spent his time finishing some oil sketches. He made one last visit to the Nek to record the effect of the dawn light, and completed the painting just before breakfast. Lambert was particularly pleased with a painting that showed his assistant, Spruce, posing like a corpse at Johnston's Jolly by lying face down on the ground. Ever determined to get the details right, Lambert commented that it was 'quite a good correct study of Spruce, the light horseman, as a stiff'. Most importantly for Lambert, he had 'the right kind of man in right clothes and right ground ... In fact everything right. A four hours' stretch and worth it!'[33]

Lambert was satisfied with the group of paintings he had produced during his time around the ANZAC area. He called them 'fair souvenirs' and 'most useful reference notes'.[34] He had worked hard to produce accurate and evocative images of places associated with Australian action. At the end of his tour he felt he knew the character and detail of the landscape and was confident that his paintings would provide him with enough material for the larger commissions that would follow. And yet the sad ambience of the place also affected him. While he was getting ready to leave the camp, he wrote:

The worst feature of this after battle work is that the silent hills & valleys sit stern, unmoved, callous of the human and busy only in growing bush, and sliding earth to hide the scars left by the war disease. Perhaps it is as well we are pulling out tomorrow, this place gives one the blues though it is very beautiful.[35]

The last task for the Historical Mission on the peninsula was a visit to the southern battlefields 'to follow the tracks of the Australians at Helles'.[36] On the morning of 7 March 1919 Bean, Lambert, Wilkins and Herbert Buchanan rode south, leaving the other members of the party to pack up the camp and take the salvaged objects down to the coast, ready to ship back to Australia.[37] James and Swanston remained at ANZAC to complete their photographic survey.

Using existing tracks, Bean led the Historical Mission down onto the Kilid Bahr Plateau; from there they could see clearly the peaked hill of Achi Baba. From its heights there was a commanding view of all the southern part of the peninsula, including the small town of Krithia. Achi Baba had been the primary military objective in the British and French campaign at Cape Helles, one they had expected to win on the first day. In the end, it was never taken. The party continued and rode down to the deserted 'V' Beach at Cape Helles, where the hulk of the old collier *River Clyde* still lay in shallow waters. The Historical Mission made its camp on a terrace overlooking the entrance to the Dardanelles. They found it a picturesque site with the sweep of the bay framed by beautiful cypresses. They spent the next two days conducting a thorough investigation of the battlefields at Cape Helles.

The land the Historical Mission covered here was very different from that of the northern ANZAC area. It was open heath or farming land with fruit and olive trees lining the roads and undulating fields of wheat. In spring time the area would be covered with flowers, including poppies, lupins and daisies.[38] The land rose gently from the coast towards Krithia and then on to the squat peak of Achi Baba about

When the weather was too poor for photography, James and his party helped the Graves Registration Unit clean up and fence the cemeteries and place some of the wooden crosses. This photograph was probably taken the day before James and Swanston left Gallipoli.

William SWANSTON (1881–unknown) and
William JAMES (1887–1972)
Brown's Dip No. 1 and No. 2 cemeteries, looking west, No. 2 the nearest, after the crosses had been erected by the Graves Registration Unit, c. 5 April 1919
Glass 10" x 12" negative
AWM G01849

Lambert painted this small study at sunrise 'to get the effect of light for the Charge at the Nek'. He found it 'very cold bleak & lonely. The Jackals, damn them, were chorusing their hate, the bones showed up white even in the faint dawn and I felt rotten, but as soon as I got to my sport, the colour of the dawn on this shrubby scrubby hilly-land was very beautiful and I did my little sketch quite well before breakfast.'

George LAMBERT (1873–1930)
Lone Pine, looking towards the Nek, Walker's Ridge, Turkey, 6 March 1919
Oil on wood panel, 13.8 x 21.8 cm
AWM ART02826

On 24 February 1919 Lambert wrote, 'I footed it to a very interesting Turkish trench on a hill called 'Johnson's Jollie [sic]' and there did quite a good correct study of Spruce, the light horseman, as a stiff. It was quite exciting in that I had the right kind of man in right clothes and right ground. In addition to correct surroundings & light I may mention the equipment – webbing equipment. In fact everything right. A four hours' stretch and worth it!'

George LAMBERT (1873–1930)
Study for dead trooper and detail of Turkish trench, Gallipoli (Pro patria),
Turkey, 24 February and 3–4 March 1919
Oil on canvas, 35.7 x 45.8 cm
AWM ART02857

Several wild olive trees marked the position of Tommies' Trench, where a temporary headquarters had been set up during the second battle of Krithia in May 1915. Bean used the trees as a reference point from which he could measure out distances.

Charles BEAN (1879–1968)
The olive tree near Tommies' Trench,
c. 8–9 March 1919
Nitrate negative
AWM G02115

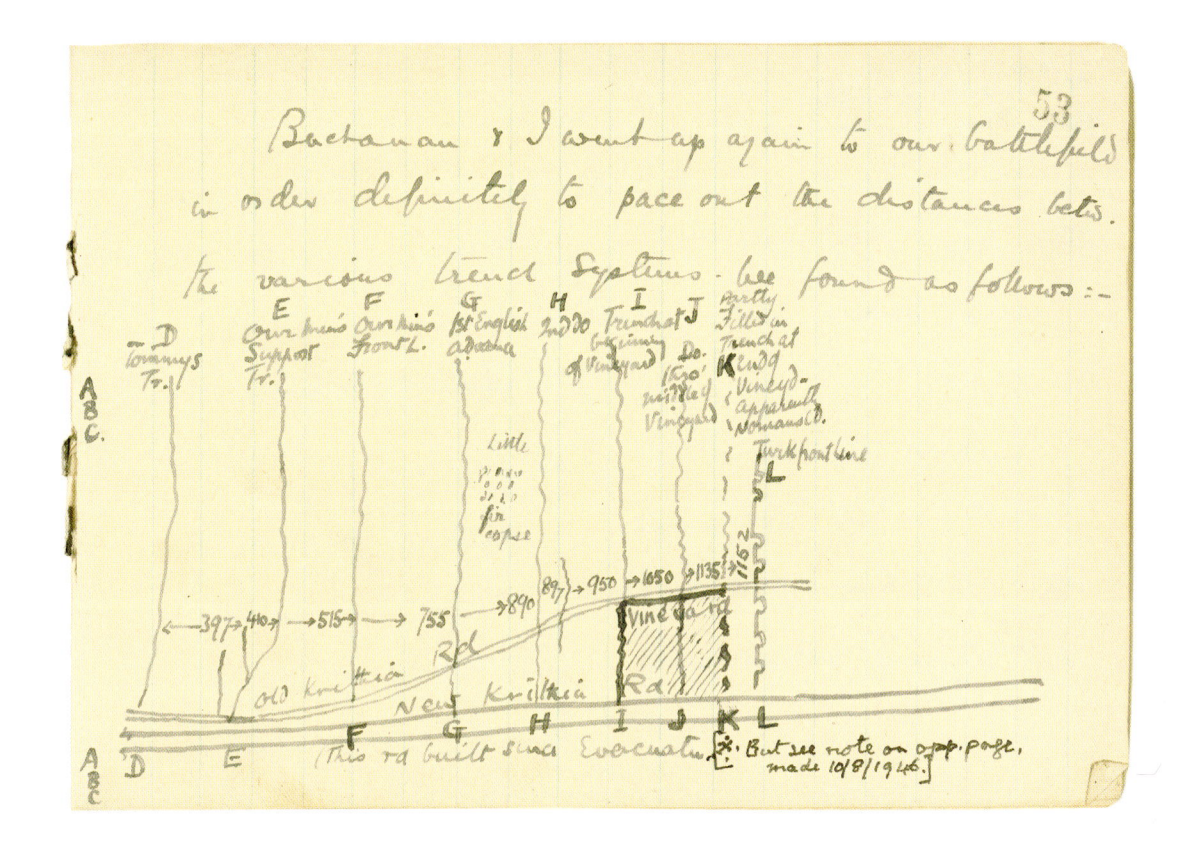

8 kilometres away. There were many excellent vantage points from which the party could survey the whole area and gain an understanding of the part the Australians had played in the second battle of Krithia in May 1915.

Bean felt a deep connection to the Helles battlefields. In early May 1915 he had gone with Australian and New Zealand troops when they were hastily transferred to the south to reinforce British and French forces in a big push towards Achi Baba. As a result he was involved in an advance made by the Australians towards Krithia on the afternoon of 8 May 'which in an hour or two, cost our brigade there (the 2nd) over 1,000 of its 2,900 officers and men'.[39] Bean had not just witnessed this carnage, he had also spent much of the day and night running to and from the front line, taking and delivering messages, and helping wounded soldiers, all the time making notes of what was happening. He later recalled: 'I knew that battlefield better than any in the war – except perhaps Pozières in France – having made the double journey

Charles Bean (1879–1968)
Bean's sketch of the distances between trenches at Tommies' Trench, 8 March 1919
Pen and ink sketch in diary no. 231, folio 53
AWM38: 3DRL/606/231

The landscape around Cape Helles is different to that of the ANZAC area. Lambert noted, 'The plain is covered with bushes of heather and sage-brush with a few small thorn trees and olives, with one or two only of short pine. The whole landscape is a dull mauvey grey with a sage green admixture and very delicate if sombre in tone. The dead, or rather their bones, spoil it, of course, and the melancholy is ready for him who lets his thoughts wander.'

George LAMBERT (1873–1930)
Achi Baba, from Tommy's [*sic*] Trench, Helles,
Turkey, 8 March 1919
Oil with pencil on wood panel, 20.6 x 29.7 cm
AWM ART02849

BATTLEFIELD LANDSCAPES: THE WORK OF JAMES AND SWANSTON

Charles BEAN (1879–1968)
The Vineyard at Cape Helles, which was part of the enemy's trench line during the attack on Krithia by the 2nd Brigade on 8 May 1915, c. 8–9 March 1919
Nitrate negative
AWM G02116

over it, up and back, four times during the afternoon and night of the advance, and once again the next day, and once more a few months later.'[40]

Although Bean devoted a whole chapter in the official history to the Australian involvement in the second battle of Krithia, he assumed the detached view of an historian and did not directly mention his own involvement. At one point, in a footnote, he modestly described an incident when a 'soldier' on the way to the front line shared a meagre amount of water with the wounded.[41] He was that soldier. He had been recommended for a bravery award for this and a number of other incidents in which he had protected and assisted the Australian wounded. As a civilian on the battlefield, he was ineligible for a decoration. However, he was Mentioned in Despatches and his bravery, his steadiness under fire and his gallantry were all remarked upon.[42]

This place had intense personal associations for Bean and it is therefore not surprising that most of the time the Historical Mission spent at Cape Helles in 1919 was devoted to clarifying details of the 1915 action in which he had been a participant. Now, with time to check and recheck details, the Historical Mission spent the best part of two days tracing the course taken by the 2nd Australian Infantry Brigade when they had been hastily thrown into the late afternoon attack.

After Bean described the event to Lambert, the artist sat down and began a painting. He positioned himself at Tommies' Trench, one of the principal places related to Bean's own story and the Australian advance on Krithia. From here, Lambert had a clear view over the forward trenches to Achi Baba in the distance. The almost flat land gave him an opportunity to spend more time developing a sense of distance and atmosphere, picking out the key features of the trenches and low-lying shrub in his study. The colours he used shifted to soft greens and mauves to best represent the different vegetation, and he left much of the wood panel unpainted to indicate the earthy soil colour of the southern region. While there is no evidence to suggest that Bean ever directed Lambert to paint a particular subject, it is clear that the artist was influenced by the historian's interpretation and the emphases he placed on certain events. As well as being a place of significant Australian casualties, Tommies' Trench had personal significance for Bean, and it is likely that Lambert wanted to commemorate this in a painting.

While Lambert continued his painting, the rest of the Historical Mission examined the area. Wilkins took photographs of the flat plain, and several of these show Lambert sitting on the edge of a trench painting his study of Achi Baba. Bean paced the distance between the trenches several times as he had a strong need to confirm the observations he had made while under intense pressure in 1915, and wanted to get as accurate a record of the ground as possible. He recorded his detailed findings in his notebook. He also spent time noting where the Turkish positions had been and what they could have seen of the attacking troops.

The need to verify these details continued to consume Bean and the next day he stepped out the distances and relationships between the trench lines two or three more times. The party also noted down possible Australian graves for the Graves Registration Unit. After this, the only thing left to do was to ascend Achi Baba and survey the battlefield in order to ascertain what the Turks could have seen from their positions. So on the morning of their departure from the peninsula, Wilkins, Buchanan and Bean made this journey to the peak, where they had a commanding view of the whole foot of the peninsula. They could see some of the features of the ANZAC area, such as Hell Spit, Battleship Hill and Chunuk Bair, but little else. Apart from a glimpse of Chanak, any view looking eastwards was largely cut off by the big shoulder of Kilid Bahr Plateau.

While Wilkins, Buchanan and Bean were at Achi Baba, Lambert stayed at Helles to work on a small painting of the *River Clyde* aground in the harbour. Everyone met up later in the day at the Kilid Bahr jetty and prepared to board the small steamer that would take them to Constantinople.

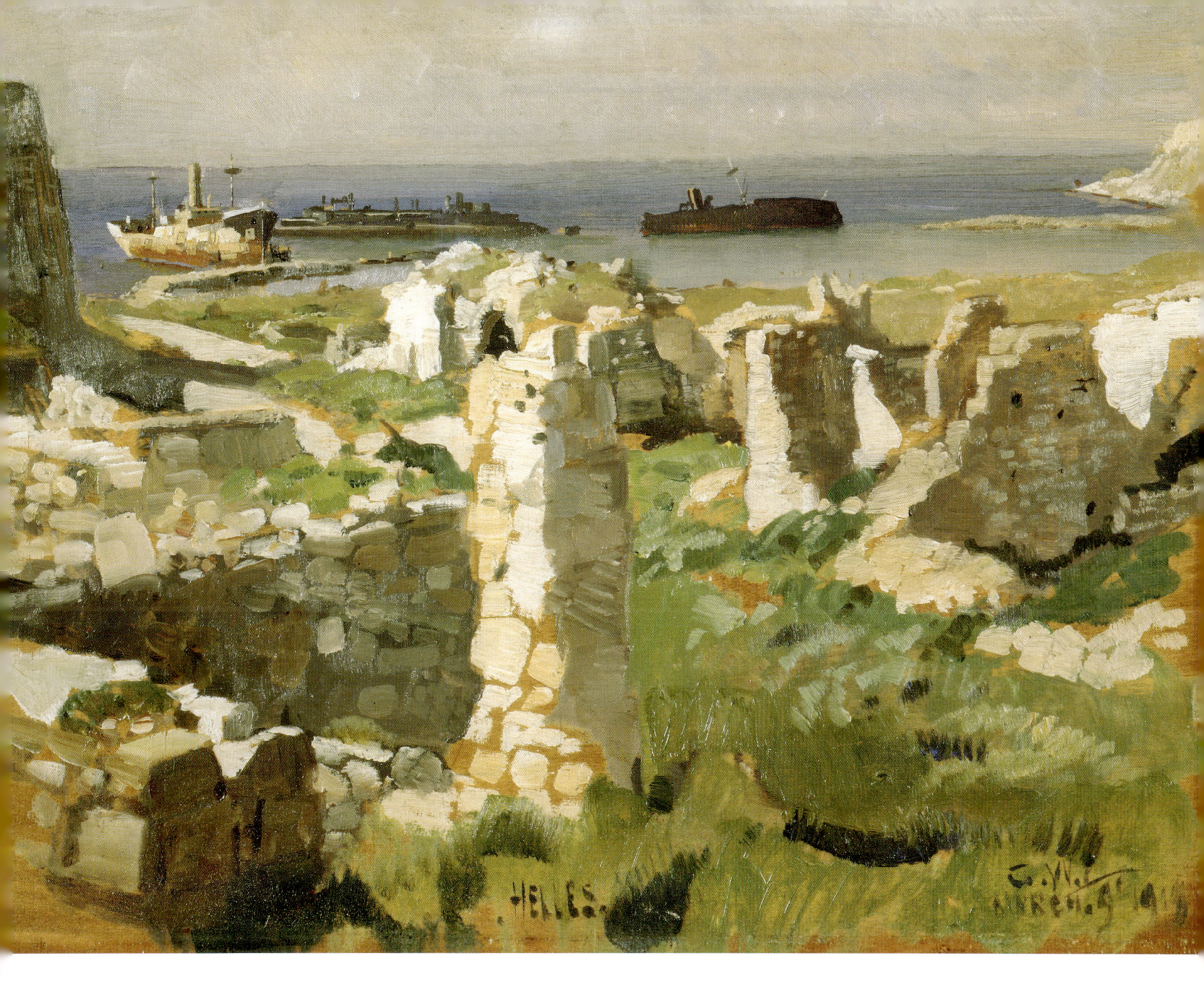

This was Lambert's final oil sketch made on Gallipoli: 'I cannot tell you how pleased I am at getting clear of this graveyard beautiful as it is nor can I explain how satisfied I am to have done what work I have done.'

George LAMBERT (1873–1930)
River Clyde at Cape Helles, Turkey, 9 March 1919
Oil on wood panel, 23.1 x 30.2 cm
AWM ART02850

Back row: John Balfour (left), Hubert Wilkins; middle row, sitting: George Rogers (left), George Lambert; front row, standing: Lieutenant Mackinnon (left) and Herbert Buchanan. Mackinnon, a British officer, was in charge of a small post where Turks surrendered their weapons.

Charles BEAN (1879–1968)
Five members of the Australian Historical Mission with Lieutenant Mackinnon, Railway Control Officer near Eregli in one of the railway trucks which served as their living quarters while travelling from Constantinople to Cairo, c. 16 March 1919
Nitrate negative
AWM G02138

CHAPTER SEVEN

THE JOURNEY HOME

IN THE MORNING OF 10 MARCH 1919, THE AUSTRALIAN Historical Mission stepped off the Gallipoli peninsula and boarded the Greek steamer *Spetsai*, bound for Constantinople.[1] There was a general feeling of release at leaving the peninsula. George Lambert described how there was a 'holiday atmosphere in the air' and he found himself 'singing in grand opera style to an accompaniment of raucous Lancashire Tommies who also felt a certain sense of relief in getting away from the graveyard work'.[2]

As the *Spetsai* pulled away, the men of the Historical Mission waved goodbye to their colleagues who remained behind. Cyril Hughes's work in preparing the battlefields and cemeteries for visitors was to continue for several years, in conjunction with the Imperial War Graves Commission. In 1919 the area was still off limits to civilians, who were turned away with a firm hand. However, compassion for mourning relatives sometimes got the better of the authorities, and within a year Hughes would kindly look after at least one father who had travelled from Australia to search for his son's grave.[3]

The AWRS team led by William James remained on Gallipoli for another month, packing the relics and awaiting transport back to Cairo. Large objects also awaited shipment, including water carts and gun limbers, one of which had been dug out of the roadside at Casualty Corner. Two lifeboats and a large gun were left behind in the hope that Hughes would be able to arrange transport for them at a later date.[4]

Once in Constantinople Charles Bean wanted the Historical Mission to take the earliest available sea transport back to Egypt. But there was another possibility. A recently opened overland railway line through the Taurus Mountains provided a link to the railways in Syria, Palestine and Egypt. The Taurus railway had been constructed as a direct rail link for German and Turkish troops moving across Asia Minor, but with the Armistice the British had taken control of the route. Most members of the Historical Mission thought a long train journey was a more attractive and leisurely way of getting to Cairo, but in Bean's opinion it would be 'purely a joy ride' and would put the Australian government to unnecessary expense.[5] In the end, Hubert Wilkins talked to British friends in charge of railway movements and was able to secure free transport in some empty livestock wagons, so they decided to take the slower journey after all.

They were allocated two enclosed wagons, and the whole group set about scrubbing them down. Lambert and George Rogers took it upon themselves to convert one of the wagons into a kitchen and eating area. They constructed a rough and ready table and some benches out of scrap wood; Bean wrote that they 'were so proud of their work – and we so appreciative of it – that they forthwith volunteered to stay there and cook for us during the trip'.[6] Throughout the 2,400-kilometre journey, Rogers, with Lambert as 'cook's orderly', served meals made from their existing supplies,

Charles BEAN (1879–1968)
Taurus railway summit,
17 March 1919
Nitrate negative
AWM G02140

THE JOURNEY HOME

Charles BEAN (1879–1968)
Glimpse of the mountains from the entrance to a Taurus Mountains railway tunnel, c. 16–17 March 1919
Nitrate negative
AWM G02145

George LAMBERT (1873–1930)
The top of the Taurus mountains,
Turkey, 17 March 1919
Oil with pencil on wood panel, 20.6 x 29.7 cm
AWM ART02849 (verso)

The Australian Historical Mission travelled through the Taurus Mountains on the way through Turkey towards Egypt. The train stopped frequently to take on water, wood and supplies. Whenever possible, Lambert would make a quick sketch of his surroundings. This one was painted on the back of one his Gallipoli paintings.

George LAMBERT (1873–1930)
Karpura, Taurus Mountains,
Turkey, c. 16–17 March 1919
Pencil on paper, 30.4 x 23.2 cm
AWM ART11393.355

supplemented by fresh vegetables and meat bought from trackside vendors. 'Stew was our masterpiece and [it was] porridge for breakfast.'[7]

Lambert frequently entertained the party with lively stories and impressions of pompous British officers he had met in Palestine. Despite his outward good humour, Lambert found that travelling in such a small party and confined space brought its own problems. He chose to sleep in the kitchen wagon as 'a little loneliness seems priceless even when one is with the right kind of people.'[8]

THROUGH THE TAURUS MOUNTAINS

Bean's diary entries virtually stopped when he left Gallipoli in March 1919. His immediate war work had ended and when he boarded the train he delivered himself into 'the care of the omnipotent railway'.[9] During the rail journey he allowed himself to relax and enjoy the sights and scenery. For Bean, this was an indulgence that had been impossible during the previous five years. 'Even in the most beautiful rest areas of France and Belgium', it was impossible for him to relax knowing that close by 'men were killing one another'.[10] For nearly two weeks members of the Historical Mission found respite from the intensity and constant anxiety of recent years. At last, it felt as if the war was really over. They could gain a glimpse into the lives of ordinary Turkish civilians and sit back and observe the spectacular landscape as they travelled through the fertile lowlands of Turkey and then up into the Taurus Mountains. As Bean poetically recorded: 'Scene after scene swam before us, drunk in with delight, but leaving the impression of a dream.'[11]

They travelled through the heartland of the country. Camera ever to hand, Bean made a few photographs during the journey. Some of these are 'tourist' photographs, taken from the train window, of the beautiful and rugged scenery.

He captured views through an almost vertical gap in the mountains and impressive vistas of the mountains covered in snow. Lambert also occasionally took time out from his cooking duties and made quick sketches. A study of the mountain summits – 'the most spectacular country in all that part of the world that I have travelled in'– was hurriedly painted on the back of one of the Gallipoli paintings while the train halted at Bozanti.[12]

This was land that had been fought over for millennia. Bean noted that it had a 'history going back farther than St. Paul'; where Assyrians, Lydians, Greeks, Romans, Egyptians, Arabs, Turks, Crusaders – and in these very modern times Napoleon and ourselves – had marched and fought'.[13] The journey was also full of reminders of the effects of the recent war and the continuing military presence in eastern Europe. As they passed the Gulf of Ismid they could see *Goeben*, the German battle cruiser that had shelled ANZAC positions in 1915, and when they reached the mountains they met overcrowded trains of Turkish troops who were being demobilised and repatriated to their homes across the country. Turkish soldiers 'overflowed from every carriage, out of the windows, sometimes on the steps and even the roofs'.[14] The process was being overseen by a small number of British and Indian troops, who were heavily outnumbered and inadequate to enforce the demobilisation and handing-in of weapons. The confusion was immense, and at Adana Bean met old Australian friends who told him that the Turkish army was secretly exploiting the situation to infiltrate troops and weapons into the interior.[15]

Bean may have stopped taking detailed notes, but it seems that he was never really off duty. His extraordinary thoroughness in recording innumerable small details and then confirming them by careful investigation before publication is illustrated by an incident at Hadschkiri, high in the Taurus Mountains. When the train stopped to take on water,

Charles BEAN (1879–1968)
The railway station near Islahiye on the Taurus Mountains railway, c. 16–17 March 1919
Nitrate negative
AWM G02149

THE JOURNEY HOME

Charles Bean's informal photographs taken from the train indicate the large number of Turkish troops being transported across the country during the demobilisation process.

Charles BEAN (1879–1968)
Arab and Turkish soldiers at Islahiye Station on the Taurus Mountains railway line, c. 16–17 March 1919
Nitrate negative
AWM G02150

Charles BEAN (1879–1968)
Mount Lebanon from the valley between Lebanon and Anti-Lebanon, c. 23 March, 1919
Nitrate negative
AWM G02156

Charles Bean had always been interested in classical history; the ruins and 'lovely pillars' of the Roman temples at Baalbek reminded him of the thousands of years of continuous human history in the land, which looked dry and dead and 'as though adders and scorpions lurked in every crevice'.

Charles BEAN (1879–1968)
The ruins of Baalbek from the railway,
23 March 1919
Nitrate negative
AWM G02159

As the Historical Mission travelled towards Cairo they frequently saw trains going in the opposite direction, carrying thousands of Turkish troops. These trains were desperately overcrowded with men perched on the steps and roofs. Deaths from sickness and railway accidents were common.

Charles BEAN (1879–1968)
Turkish troops being crowded into trucks for passage through the Taurus Mountains tunnels at Adana, c. 18–21 March 1919
Nitrate negative
AWM G02134

Charles BEAN (1879–1968)
Turkish troops leaving a train in the Taurus Mountains, c. 16–17 March 1919
Nitrate negative
AWM G02151

Charles BEAN (1879–1968)
The daily train being pulled by a German engine on the Taurus railway line at Karpura, c. 16–17 March 1919
Nitrate negative
AWM G02143

he searched for the graves of the Australian prisoners of war who had died while put to work constructing the railway. This was another of those riddles he felt compelled to unravel. He scrambled a couple of hundred metres up the hill and 'there, derelict and overgrown with grass, were a dozen or twenty graves' of Australians, Britons and New Zealanders.[16] As these isolated graves were unlikely ever to be visited by relatives, Bean took photographs that he later passed on to the families.

One of these was the Calcutt family who had lost a son, Gerald, on Gallipoli in May 1915.[17] Their second son, Brendan, was wounded in August 1915 but saved from immediate death by a German officer who prevented Turkish soldiers rolling the injured man over a cliff.[18] Held as a prisoner of war, Calcutt subsequently died working on the railways at Hadschkiri.[19] His story was told in the second volume of the official history series and in the photographic volume; one of Bean's photographs was used as an illustration of typical workers' huts where the Australian prisoners of war were housed at Hadschkiri. These were the small details that completed the web of information Bean had amassed.

As they emerged from the mountains, the Historical Mission passed through landscapes that had an ancient human history. Bean often commented on the classical and biblical sites they encountered. His interest in history was constantly stimulated by places that were associated with people such as Cicero or St Paul or the Crusaders. Later, at Baalbek in eastern Lebanon, he was reminded of the more than 5,000 years of documented history in the area by tantalising glimpses of the ruins as the train made a brief stop.

The Historical Mission travelled down onto the plain and stopped at Adana. Here they left their cramped quarters on the train, visited the shops, and observed the traders in the open-air market. The market was crowded with produce-sellers as well as people providing services (letter-writing, for example) or dispensing herbal remedies. Evidence of the

Photographer unknown
A group of nine (unidentified) Australian prisoners-of-war who were employed on railway construction work at Bozanti and along the Taurus Mountain Railway, 21 June 1918
Gelatin silver print
AWM H19402

Some Australian prisoners of war captured by the Turks on Gallipoli were used between 1915 and 1918 to help build and maintain parts of this railway. Bean found and photographed the graves of Australian prisoners who died near Hadschkiri.

war was never far away, and the streets were crowded with troops of all nationalities. Bean took particular note of the French infantrymen of Armenian extraction who were responsible for protecting the Armenian population from the Turks. The Armenians themselves, however, were equally troublesome, and the British found themselves uncomfortably in the middle of a continuing conflict.

Arriving at Aleppo on 22 March, they were delighted to find an old friend, General Harry Chauvel.[20] A veteran commander of the Light Horse on Gallipoli and of the Desert Mounted Corps in Palestine, Chauvel was well known to Bean, but also to Lambert, who, as an official war artist in 1918, had been attached to his headquarters.[21] At this time Chauvel was representing the British army in Aleppo and attempting to enforce the terms of the Armistice. Bean and Chauvel discussed the difficulty of managing the retreating Turkish army. Mustafa Kemal, the Turkish leader who had played such an important part in defending Gallipoli in 1915 and who would later be instrumental in the formation of the modern Turkish state, was proving 'a stubborn and difficult man to deal with'.[22] In addition, the undisciplined Arab forces were beginning to put aside their regional and tribal differences and loosely unite under Emir Feisal, which added another layer of complexity to the postwar politics.[23]

As they left Chauvel's headquarters they passed through country that Bean thought 'looked as though adders and scorpions lurked in every crevice'.[24] Any roughing it ended for the Historical Mission when they arrived in Damascus the next day. Here they met Colonel T.E. Lawrence – already famous as Lawrence of Arabia – whom Rogers noted was 'just as interested in us as we were in him'.[25] Lawrence and his Bedouin followers offered to put on a display of horsemanship for the Historical Mission. The riders simulated charges 'waving scimitars and firing small arms, to the accompaniment of blood curdling yells and thundering hooves'.[26]

In Damascus, Lambert's role as cook's orderly ended, and the group transferred to another train that took them to Jerusalem. During the last few days of the journey they crossed landscapes that had become part of Australia's military history and where the Australian Light Horse had learned to adapt to desert conditions. They were reminded of this when, approaching Cairo, they saw groups of Australian soldiers ('laughing, confident, clean-looking youngsters', in Bean's words) who were waiting to return to Australia.[27]

THE JOURNEY HOME

Adana was predominantly an Armenian centre until 1909, when religious and ethnic tension erupted, resulting in many deaths. After the Armistice of 1918, British and allied troops were stationed in Turkish towns to keep order and prevent factional fighting among Turks and Armenians.

Charles BEAN (1879–1968)
A British patrol and demobilised Turkish soldiers at Adana, 18 March 1919
Nitrate negative
AWM G02152

George LAMBERT (1873–1930)
Aleppo citadel, Syria, 22 March 1919
Pencil on paper, 23 x 30 cm
AWM ART02869

George LAMBERT (1873–1930)
Jerusalem from the top of the Dung Gate, Palestine, 25 March 1919
Oil with pencil on wood panel, 19.6 x 45.9 cm
AWM ART02855

CAIRO DEPARTURES

The Historical Mission arrived in Cairo on 26 March, and once there, the group started to split up. They had been travelling together for a total of 68 days since leaving London. Their journey had taken them through France and Italy, across the Aegean Sea to the Gallipoli peninsula, and then on to Constantinople, through mountainous Anatolia and then down through Syria, Palestine and into Egypt. Almost ten weeks of travelling had helped to create lasting friendships and a sense of common purpose that would continue well after the expedition.

They were met by Arthur Bazley, who had recuperated from influenza and travelled to Cairo, as well as Hector Dinning from the Cairo subsection of the AWRS. Although the war was over, Dinning still managed a team of people who performed all sorts of practical, administrative and record-keeping tasks as well as facilitating the work of the official war photographers and artists in the Middle East. With Australian units now demobilised, the AWRS was keen to secure material before the units disbanded; there was also a rush to collect unit diaries and official documents. When the Historical Mission arrived, a massive project to collect relics from battlefields in Sinai, Palestine and Egypt was well underway, and the physical magnitude of the collection being amassed was already apparent. At the time, Dinning was attempting to secure a light railway engine and a large pontoon that the Turks had dragged across the desert to attempt to cross the Suez Canal.

In 1919, when the Historical Mission arrived in Cairo, the AWRS had just finished building and equipping a new dark room, which would be managed by Oswald 'Ossie' Coulson, an AFC photographer and subsequently an official photographer with the AWRS. Coulson was an excellent choice as he was an experienced lithographic artist and photographer prior to enlistment in 1916. The Historical Mission was carrying the glass-plate negatives that Wilkins had taken on Gallipoli. As a precautionary measure against damage to the glass plates, it was essential to print copies of everything before Bean departed for Australia with the precious negatives. Some of Wilkins's plates had been printed by William Swanston while on Gallipoli, but at least two complete sets of his Gallipoli photographs were made by the AWRS dark room staff in Cairo between 26 March and 2 April 1919.[28] It is probable that Coulson supervised the printing of these plates.

One of these Cairo sets was annotated by Wilkins with titles and remarks on the back of the images. This set was taken back to Australia by Bean and was later pasted into a large-format album that he used for reference while writing the official histories.[29] The other set was taken by Wilkins to London and handed over to John Treloar at the AWRS headquarters in Horseferry Road. It is not known what happened to this group of images.

On 2 April 1919 Bean, Bazley, John Balfour, Herbert Buchanan and Rogers left Cairo for Australia aboard the *Kildonan Castle*.[30] Wilkins and Lambert remained behind. Little is known of what Wilkins did while in Cairo, but it seems probable that as well as preparing his own photographs for distribution he visited his friends in the Australian and British flying units before leaving for London on 12 April.

Led by James, the AWRS group that had been left guarding the relics on Gallipoli returned to Cairo on 12 April. James had with him about 150 exposed glass plates that he and Swanston had taken on Gallipoli. The possibility of damage to the fragile negatives was always present and it was imperative to process them and print photographs as soon as possible. The AWRS dark room was running at capacity, producing photographs for departing soldiers and field units, so the RAF again stepped in and made its own facilities available. Swanston's assignment was extended to complete the photographic work for James.[31] Once James had delivered all the Gallipoli material to the AWRS offices in Cairo, he was employed for three months surveying battlefields and collecting relics. He finally departed for Britain on 23 July 1919. Swanston also left the Middle East in May and returned to Britain, where he was discharged from the RAF on 30 April 1920.[32]

An important task of clerks in the AWRS in Cairo was to copy and transcribe documents, including the unit diaries. Seen in this image are Sergeant Fawcett and Corporal Tiddiman.

Charles BEAN (1879–1968)
The war diary room at the Australian War Records Section, Cairo, 1919
Glass whole-plate negative
AWM B01391

George LAMBERT (1873–1930)
Portrait of C.E.W. Bean, left profile, probably Palestine, 1919
Pencil on paper, 17.8 x 13 cm
AWM ART90829 (gift of Mrs C.E.W. Bean, 1997)

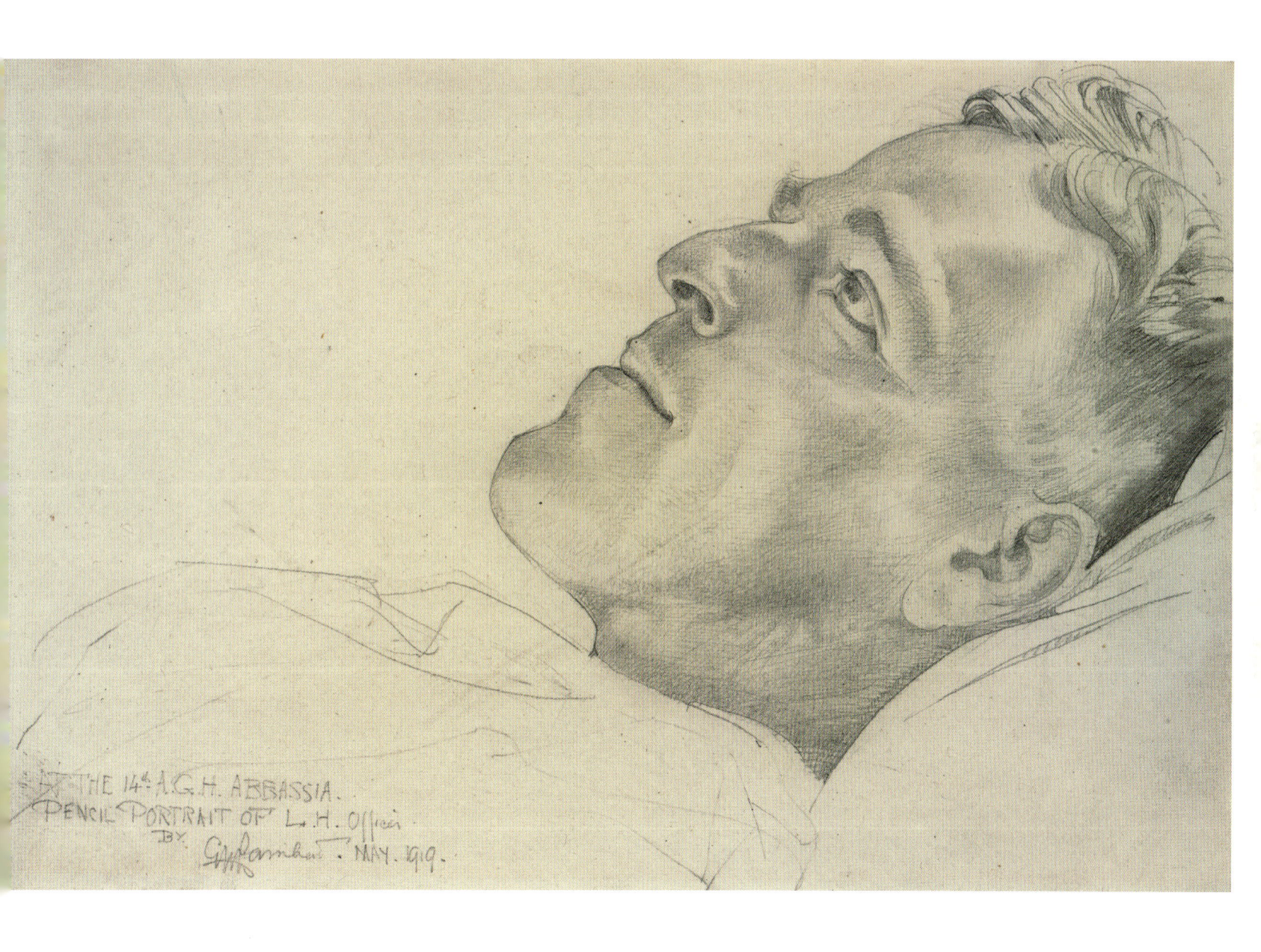

George LAMBERT (1873–1930)
Light Horse Officer in 14th Australian General Hospital, Abbassia, Egypt, April–May 1919
Pencil on paper, 16.6 x 25.2 cm
AWM ART02871

When the Historical Mission arrived at Cairo, Lambert fell seriously ill and was hospitalised for nearly eight weeks, suffering from malaria, dysentery and an undiagnosed heart problem. As he recuperated he spent some time making sketches and paintings of Australians in the wards around him.

George LAMBERT (1873–1930)
Balcony of troopers' ward, 14th Australian General Hospital, Abbassia, Egypt, April–May 1919
Oil and pencil on wood panel, 32 x 45.6 cm
AWM ART02815

Lambert had been tempted initially to take on the Gallipoli trip by an offer to extend his journey to paint in the warmth and sunshine of Palestine. There was also a suggestion that he might be able to tour some areas he had not seen before with a group of artists and modellers expected to arrive in Cairo soon. On the day Bean and his party left Cairo, however, Lambert collapsed and was hospitalised. He had indicated before leaving Britain that his health was not good, and it seems that throughout the expedition he suffered from the persistent effects of malaria, dysentery and severe exhaustion. These were a dangerous enough combination, but in Cairo an unspecified heart problem was added to the mix. Lambert was in hospital for about eight weeks. It appears that he, as well as many around him, believed that his life was threatened. At one stage he even wrote to his wife Amy with advice on how to dispose of his small estate. He blamed his collapse on the Gallipoli trip and reassured Amy: 'There is nothing hereditary about my cardiac trouble … it was brought about purely by over Athleticism, & not drink or tobacco but the latter has decidedly increased the trouble.'[33]

As he recuperated in Abbassia General Hospital on the outskirts of Cairo, Lambert was able to make several paintings of hospital life around him. One of these was an exquisitely drawn study of an unidentified Australian Light Horse officer lying in bed. The work has an almost ethereal quality; it shows the man gazing forward, lost in his own thoughts. Lambert was deeply impressed by the courage and commitment of light horsemen and observed that he would be 'perfectly willing to go with them anywhere, any time.'[34]

By the time Lambert was well enough to resume his painting tour, Bean, Wilkins and the others were long gone. After his recuperation he spent several months painting in Palestine – the reward Bean had promised him. He visited sites associated with Light Horse units, including Barada Gorge and Semakh, where he found 'pictures by the mile', but his sense of duty forced him to stick to military subjects.[35] Despite the warmth and light he found it difficult to regain his health and optimism. In July, as he waited in Moascar for transport back to London, he noted how the Australian camps were being dismantled and a sense of melancholy pervades his journal. 'Everything is closing up here & there is a beastly left behind feeling about us all.'[36] By the time he eventually embarked on the *Caledonia* for London on 2 August 1919, Lambert was the last member of the Historical Mission still remaining in the Middle East.

Photographer unknown
Interior of the Australian War Museum, Melbourne Exhibition Buildings, showing exhibits in the show cases and paintings and photographs on display, c. 25 April 1922
Glass whole-plate negative
AWM J00290

CHAPTER EIGHT

ART, PHOTOGRAPHY AND HISTORY

DURING THE THREE WEEKS THE AUSTRALIAN HISTORICAL Mission was on Gallipoli, it amassed a staggering and diverse collection of primary evidence related to the experience of Australians in 1915. William James estimated that together with Charles Bean's party, he and his men had secured about eight tons of relics for the museum collection.[1] Twenty-six ammunition boxes contained small items collected from the battlefields: strands of barbed wire, water bottles, badges, buttons, colour patches, shell cases, pieces of tunic and the like. Two cases contained larger objects and a selection of timbers that had been used at Lone Pine. Other items were also identified to be retrieved, such as two abandoned lifeboats, one of which would eventually be added to the museum collection.

The intention of Bean and the AWRS was that some of these relics would be redistributed to various state and regimental museums. And indeed, many objects were given to other collections in subsequent years. However, items that were considered crucial to the telling of the ANZAC story were retained for what became the National Collection, housed in the new Australian War Memorial in Canberra. Some of these, such as the logs from Lone Pine, the lifeboat and a number of intensely evocative personal items found in the field, still form the core of the Australian War Memorial's permanent Gallipoli displays.[2]

George Lambert's field work during the expedition produced just over 30 small paintings, watercolours and drawings. These were accessioned into the National Collection in 1921 and were included in the Memorial's first displays in Melbourne in 1922. The photographic material created during the Historical Mission was comprehensive. Judging from museum registers and the existing original glass plates, Hubert Wilkins took about 200 images in the 24 days he was on Gallipoli. He may have exhausted his supply of plates by the end of his stay as he took no further photographs during the Historical Mission's journey through the Anatolian and Taurus Mountains. The photographic work of James and William Swanston from the AWRS amounted to some 150 glass-plate negatives, some of which may have been damaged in transit as only 139 were ever formally registered.[3]

Bean's own photographs numbered 95, of which 30 were taken while the Historical Mission travelled through Turkey on its way to Egypt. As Lambert and Wilkins produced little work during this part of the journey, Bean's images provide a valuable insight into the process of demobilisation in Turkey. In addition, Bean filled seven notebooks with topographical sketch maps and detailed diary entries, and took extensive shorthand notes of interviews he conducted with Major Zeki and other veterans of the Gallipoli campaign. Maps made on

William JAMES (1887–1972)
Members of the Australian War Records section and the 7th Light Horse Regiment at Kilid Bahr with wagons filled with war material collected on the Gallipoli peninsula, March 1919
Gelatin silver nitrate
AWM P07906.048

the spot by Herbert Buchanan and George Rogers, and those gathered from other sources and checked against the landscape, added another layer of intricate knowledge of the land across which Australian soldiers had fought.

The relics and paintings, all of the photographic images, the annotated field maps and Bean's documentary material became part of the evidence collected by the Historical Mission, now preserved by the Memorial. As a body of primary source material compiled so soon after the campaign, it has been extensively used since 1919 to help people understand and interpret the Gallipoli story.

THE OFFICIAL HISTORIES

When Bean left Cairo for Australia in April 1919, he had already begun preparatory work for the official histories. During the voyage he started to sort and index copies of official documents, interviews and personal stories submitted by soldiers, and to organise his war diaries and notebooks from the last five years. By October that year Bean and his staff were established in their new headquarters, the government-managed Tuggeranong homestead on the Murrumbidgee River plain, just south of Canberra.

From left to right: Corporal Robinson, draughtsman; Private Brown, draughtsman; Staff Sergeant Bazley, librarian; and Sergeant Rogers, draughtsman.

Charles BEAN (1879–1968)
Four members of the staff of the *Official History of the AIF* on HMT *Kildonan Castle*, at sea, April 1919
Nitrate negative
AWM G02154

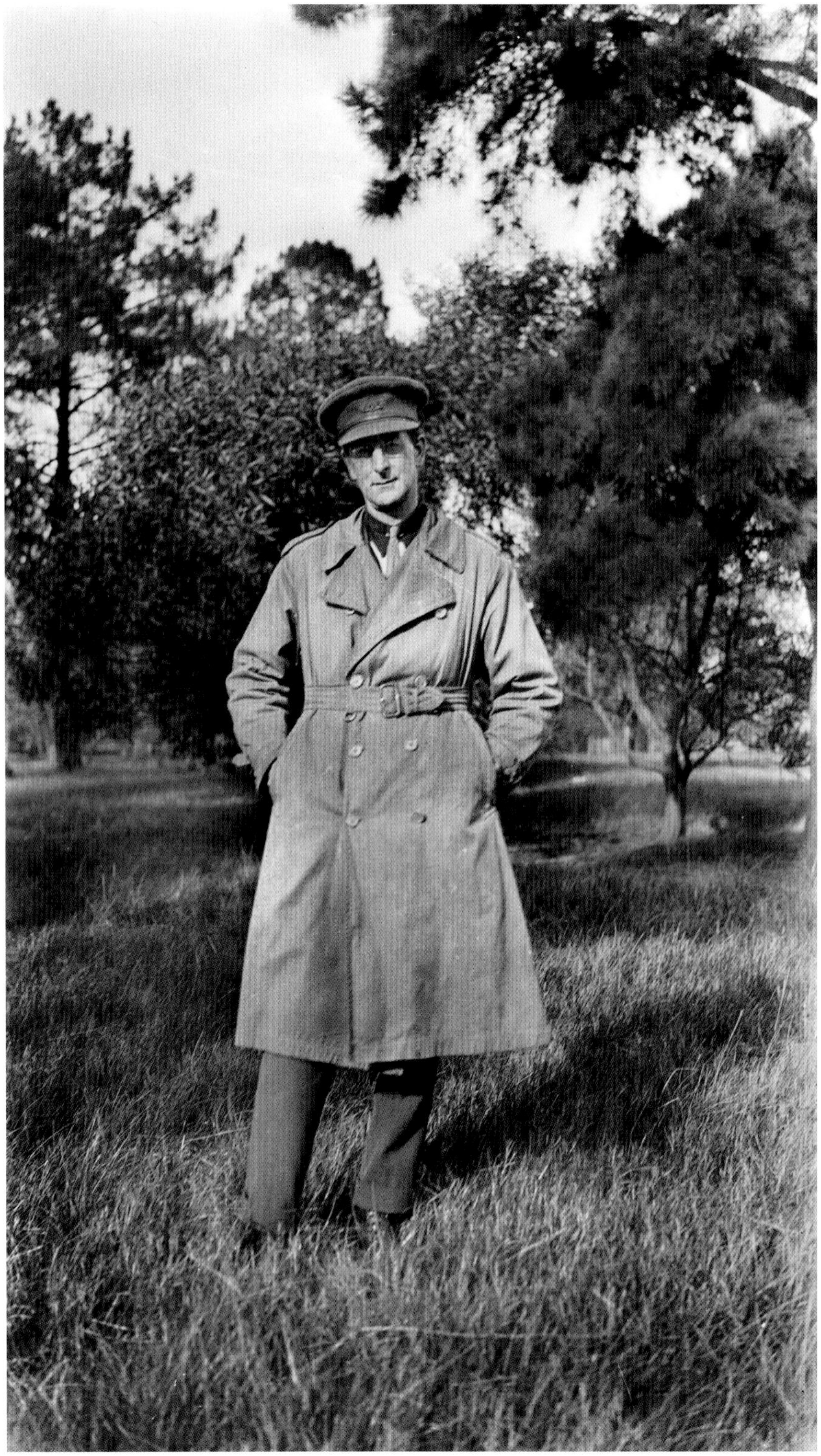

Bean had considered it appropriate that the history of Australia's part in the war should be written in the national capital and he found Tuggeranong homestead to be charmingly situated 'away by itself over the hills from Canberra in a valley of its own'.[4] The 18-room house was surrounded by 60 acres of land; it had a few neighbours but little passing traffic. The presence of the Royal Military College at Duntroon meant that military people visited Canberra, but Tuggeranong was not close enough to entice too many casual visitors. The rural setting provided the perfect location for the serious task Bean and his team were undertaking, and they were able to work uninterrupted on their correspondence, indexing, compiling, writing and fact-checking.

Bean retained the services of his two principal assistants, John Balfour and Arthur Bazley. Balfour was Bean's 'backstop', checking all the facts and compiling the short biographical notes on every serviceman mentioned in the text. The notes are a feature of the official histories and acknowledge the lives of thousands of Australians. Bazley described himself as the librarian and 'deviller', as he paid attention to the smallest details, kept all the papers in order and maintained an index of the documents so that they could quickly locate a reference to a person, place or event.[5] He also solved problems, typed up material, ran errands and generally supported Bean in his work. Other staff came and went over the years, but Bazley and Balfour remained vital to the efficient organisation of the team.

Bean surrounded himself with an enormous amount of material related to Australia's part in the war, and the range of sources he used is remarkable. He had access to the unit war diaries, official orders, signals and instructions, maps and aerial photographs, correspondence, reports, eyewitness accounts, personal diaries and signed statements, not to mention his own diaries and notebooks. He acquired copies of British, French, American and Canadian records and, wherever possible, gathered accounts from Turkish and German authorities. He consulted with many people, crosschecked oral and written accounts, and sought clarification on the smallest of details. He stated that as a historian he adopted legal rules of evidence and discarded hearsay.[6] As far as possible he only gathered evidence 'from those who actually saw and took part in the particular events narrated', but within these constraints he was aware that eyewitness accounts of the same episode rarely agreed and that a person's memory might shift or be shaped by various other factors over time.[7] He used his notes and diary entries as a starting point but also admitted that his own account was fallible and needed to be corroborated by other evidence.[8]

Arthur BAZLEY (1896–1972)
Informal portrait of journalist, historian and Australia's official war correspondent during the First World War, Captain Charles Edwin Woodrow (C.E.W.) Bean, standing in a paddock near Tuggeranong Homestead, 1919
Nitrate negative
AWM P03743.009

Photographer unknown
C.E.W. Bean and his wife Effie at the homestead, Tuggeranong, April 1921
Nitrate negative
AWM A05398

ART, PHOTOGRAPHY AND HISTORY

Photographer unknown
The first volume of the *Official History* packed and ready for transport from Tuggeranong to the publishers, Angus & Robertson, in Sydney, NSW, 12 December 1920
Copy negative
AWM A05383

Surrounded by official files, maps, photographs and reference books, Charles Bean assiduously checked and rechecked details to make the official histories as accurate as possible.

Photographer unknown
C.E.W. Bean working on official files in his Victoria Barracks office, Sydney, during the writing of the official history, c. 1935
Copy negative
AWM A05389

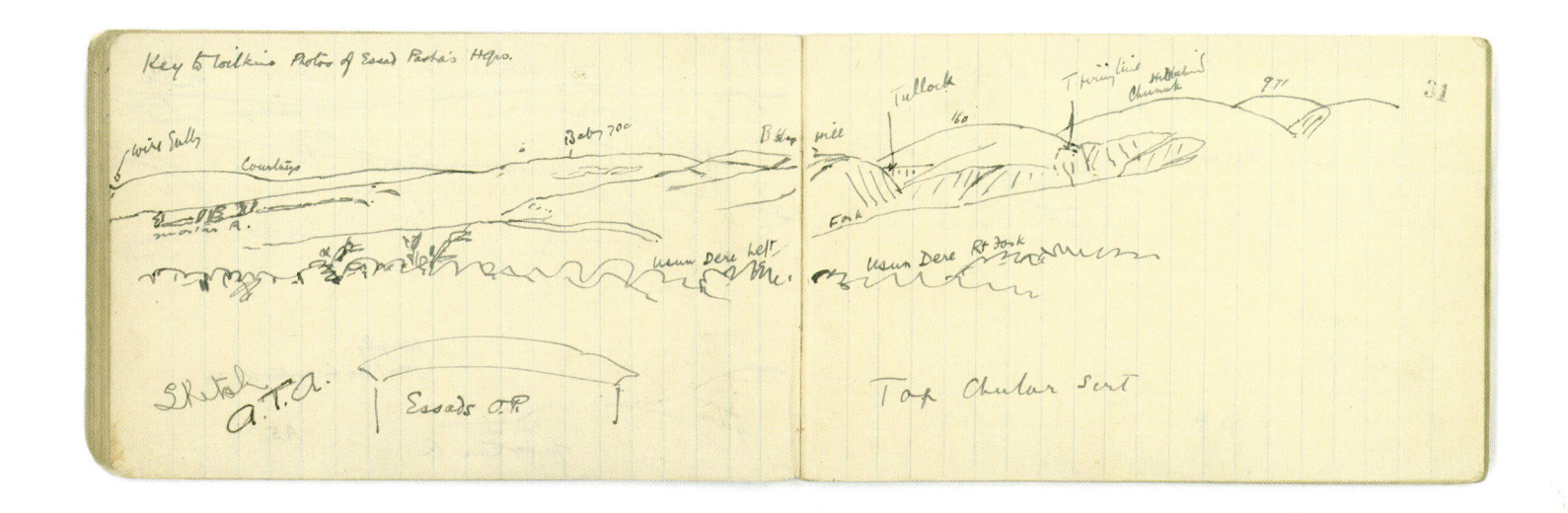

Standing alongside as Wilkins set up his camera, Bean made this quick topographic sketch of the view that the commander of the Turkish III Corps, Essad Pasha, had from Third (or Gun) Ridge across to the ANZAC positions.

Charles BEAN (1879–1968)
Key to Wilkins's photographs of Essad Pasha's Headquarters, Turkey, 6 March 1919
Pen and ink sketch in diary no. 231, folio 30–31
AWM38: 3DRL/606/231

PHOTOGRAPHS AND HISTORIES

Bean was keen that the histories he wrote should be generously illustrated with photographs. Throughout the life of the official photography scheme he had done his best to ensure that these were reliable and truthful documents. However, the scheme had not existed during the Gallipoli campaign, so for the first volume of the history, dealing with events on Gallipoli up to 4 May 1915, he used many of the photographs he had taken that year. When he needed something else to demonstrate a particular point, he turned to those made during the Historical Mission in 1919.[9] It is clear that Bean had great faith in the work of Wilkins and used 17 of his photographs as illustrations in the first volume; he formally acknowledged Wilkins's efforts 'in procuring a scrupulously accurate photographic record'.[10]

Bean was very particular about how he arranged his material, and most of it was indexed and cross-referenced. The Gallipoli photographs that he had carried from Cairo already included Wilkins's handwritten annotations to enable easy identification. Back in Australia, these photographs were pasted into large albums and further worked on by Bean. With scientific precision, he plotted key geographic features, angles and distances, and the location of specific units and military objectives across the surfaces of the photographs. As he wrote, he used these albums as primary reference sources.[11]

This point can be illustrated by studying how Bean used one of Wilkins's photographs to examine the question of precisely how far the Australian troops had advanced on the first day. This was an important riddle to solve, so Bean mapped out on a photograph the course of Lieutenant Noel Loutit's party on Gun Ridge on 25 April 1915.[12] Gun Ridge was an area close to the Historical Mission's base camp in Legge Valley and the group got to know it well. They walked it several times, making topographic maps and sketches, taking photographs, and examining the ground for evidence.

A number of relics collected from this area helped establish the position of the farthest point reached by Australians. While Bean had not been on Gun Ridge in 1915, he had stood alongside Wilkins in 1919 when he took a photograph looking from Johnston's Jolly across to the Ridge. For Bean, seated at his desk at Tuggeranong a year later, the photograph acted as an *aide memoire*, helping him to recall his own experiences on the peninsula. Looking at it doubtless brought back to him the time spent walking the ground, but it also served a more objective role, allowing him to check specific details.

Bean knew that his written version of this event must be congruent with the geography of the landscape, so he checked eyewitness statements against the photograph for consistency. The image provided visual evidence of the ground over which Loutit's party of a handful of men had advanced inland. Bean marked the photograph with the farthest point they reached. Other key landscape features, such as Lone Pine and Owen's Gully, were also marked to help make connections within a slightly broader context. In this way, he used the photograph as a framework on which to reconstruct the story. When annotated like this, the photograph summarises the trajectory of the Australians across the landscape, showing their immense achievement and their utter isolation from the main forces.

This annotated photograph was reproduced in Volume one of the official history to accompany the story of the Australian 3rd Brigade on 400 Plateau on the day of the landing.[13] Despite Bean's almost forensic approach to the detail of this event, he could only work with the information available at the time, and in the third edition of the history, he noted that new evidence indicated that two scouts of the 10th Battalion – Private Arthur Blackburn and Lance Corporal Philip Robin – may well have gone beyond Loutit's position on 25 April.[14]

The history of the AWRS photographs is slightly different. Their primary function was to document with almost scientific precision the topography of Gallipoli. From these images, craftsmen could then make three-dimensional scale models that would be used in the museum displays.

The 'plan models', as they were called, were made in the early 1920s to help people who had never been to Gallipoli understand the terrain and the geographical and strategic problems facing the troops. One plan model of the ANZAC area remains on display in the Memorial's Gallipoli gallery and gives some insight into the care and skill that was applied to ensure its accuracy.[15] The use of the AWRS photographs as reference material for the modellers may have meant that the images were not immediately available to Bean, as none were used to illustrate the first volume (published in 1921). By 1924, however, when the second volume was published, there were ten AWRS photographs reproduced.

The inclusion of Historical Mission photographs in the official histories was just one use for this material. In the years immediately after the war, the Memorial's vast store of photographic images was made available for purchase to the general public. The sales section offered a range of services including options to buy enlargements of black and white, sepia or hand-coloured photographs, and the scheme was immensely successful. It was promoted through special exhibitions that predated more permanent displays, such as that organised by the Memorial at the Melbourne Aquarium in 1921. Aware that some sections of the community might be concerned about the potentially graphic nature of the images, the Memorial assured the public that 'nothing gruesome will be shown' and offered at the Aquarium a special children's program of performing animals, magic and musical entertainments.[16] A selection of 263 photographic prints representing 'the work of the A.I.F. on Gallipoli, and in France, Belgium and Palestine' was displayed, but this represented only a small portion of the total collection.[17] Sales staff helped viewers locate other photographs that might be relevant and arranged framing and freight. In five weeks 82,998 people visited this exhibition, and the sales section was kept busy with orders for months.

Although the cover of the exhibition catalogue proudly claimed that all AIF actions would be well represented, there were in fact only two Gallipoli photographs noted as being on display. One was a scene in Shrapnel Gully taken by Bean in 1915, the other a panoramic view of Ocean Beach taken by

ART, PHOTOGRAPHY AND HISTORY

On his return from Gallipoli, Charles Bean had Wilkins's photographs pasted into a large format album. He annotated many of these as part of the process of understanding and illustrating details of the Gallipoli campaign. This page uses a photograph by Wilkins to illustrate Australian positions.

Charles BEAN (1879–1968)
Annotated page from photograph album, c. 1919–23
AWM38: 3DRL 6673/1016 (detail)

View from B (on map) looking in direction of dotted red arrow.

Lone Pine.

head of Owens Gully
Owens Gully
Johnson's Jolly
wire gully behind J. Jolly
Owens Gully Red (dot) arrow.
Foot of wire gully

Bean used a Wilkins photograph (G01899) to mark out Australian positions.

Charles BEAN (1879–1968)
Annotated page from photograph album, c. 1919–23
AWM38: 3DRL 6673/1016

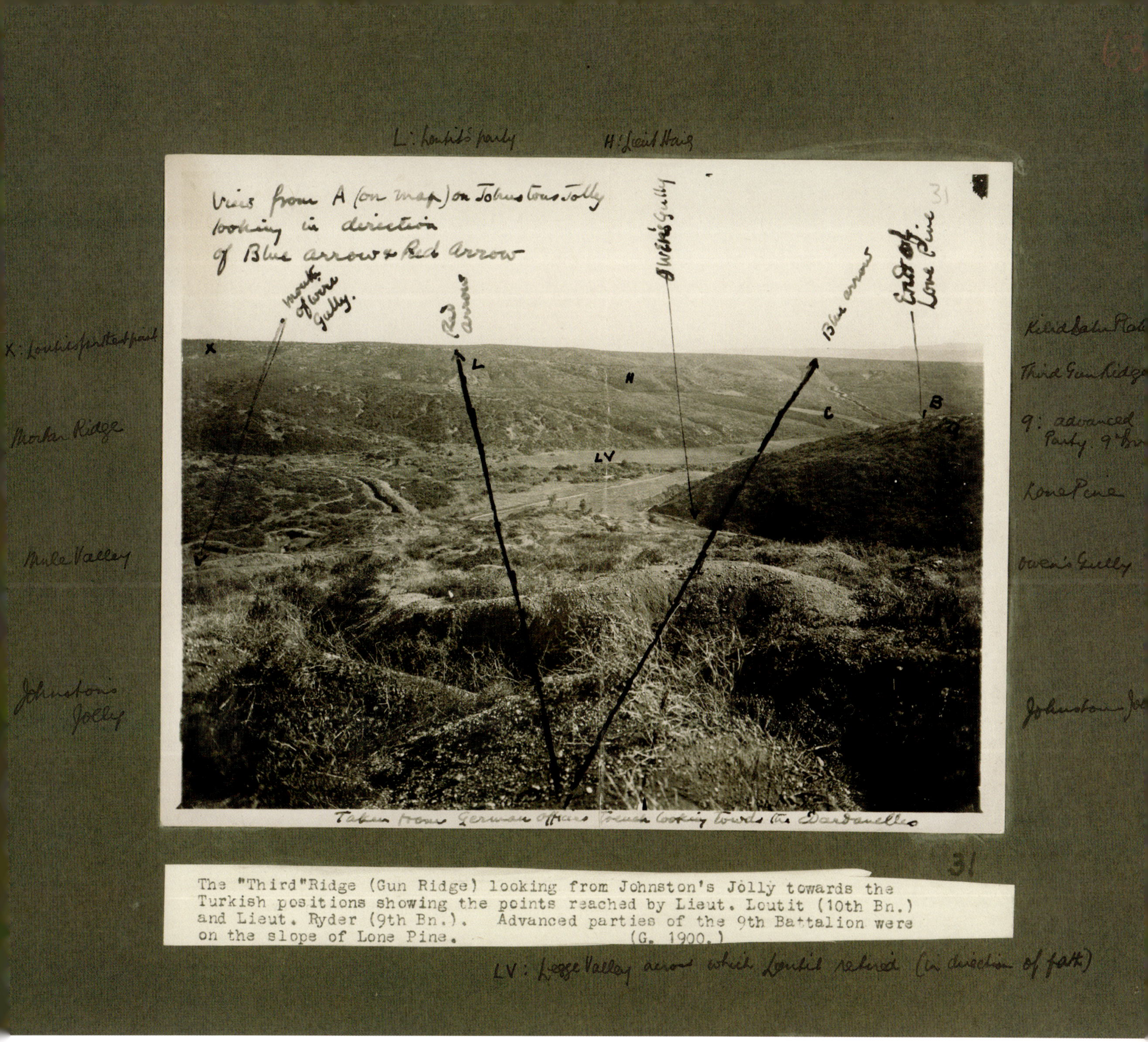
L: Loutit's party
H: Lieut Haig
View from A (on map) on Johnston's Jolly looking in direction of Blue arrow & Red arrow
Red arrow
Blue arrow
End of Lone Pine
Mortar Ridge
Mule Valley
Johnston's Jolly
Third Gun Ridge
Lone Pine
Owen's Gully
Taken from German Officers Trench looking towards the Dardanelles
The "Third"Ridge (Gun Ridge) looking from Johnston's Jolly towards the Turkish positions showing the points reached by Lieut. Loutit (10th Bn.) and Lieut. Ryder (9th Bn.). Advanced parties of the 9th Battalion were on the slope of Lone Pine. (G. 1900.)
31
LV: Legge Valley across which Loutit retired (in direction of path)

The photograph used on this page (G01900) demonstrates the path of Lieutenant Loutit and his party on 25 April 1915. The annotated image was also later used in volume one of the official history.

Charles BEAN (1879–1968)
Annotated page from photograph album, c. 1919–23
AWM38: 3DRL 6673/1016

Wilkins in 1919.[18] The scant representation of Gallipoli in this 1921 exhibition is curious: perhaps it can be explained by the fact that the first volume of the official history was about to be launched and a further volume devoted purely to photographic images was already in preparation.[19] The sales section may have been mindful of protecting its own commercial interests. However, it is also clear from Bean's own use of the photographs, and from the various guide books that were produced to accompany the Memorial's early displays, that photographic images were not generally intended for public display. Their primary function was as research documents or illustrations to support military history publications. This is how Bean used photographs within his own work, and it is one reason why he was so insistent on veracity in the work of the official photographers.

The early emphasis on official photographs as objective visual documents meant that the AWRS placed great importance on recording where and when an image was taken and what it depicted. This was crucial to the future use of the images as research documents. The attribution to a specific photographer was less important, and most official photographs entered the collection without this detail. In later years this has contributed to difficulties in identifying the work of individual Australian official photographers who worked across the Western Front from 1917 onwards.

Indeed, the collection of photographs made by James and Swanston on Gallipoli in 1919 remained attributed merely to 'AWRS photographers' until very recently, when it was confirmed that the former had been the leader of the party. Biographical details for Swanston, however, remained unknown until recent research uncovered his prior history as a professional photographer. As we have seen, this knowledge of Swanston's background makes it more likely that it was in fact he who took the images now in the Memorial's collection. Uncovering such details has been the result of increased curatorial interest in the role of the photographer as an artistic creator. This focus on the 'maker' has ensured that research will continue to strive to establish authorship for many of the Memorial's official photographs.

ART AND THE 'GALLIPOLI SPIRIT'

On 25 April 1922, when the Memorial opened its extensive displays at the Exhibition Buildings in Melbourne, only a few of the tens of thousands of official photographs were displayed. However, works of art produced by official and non-official war artists were very prominent, with one newspaper even suggesting that there was little space left for photographs as the paintings were so numerous.[20] Paintings, watercolours and drawings lined the walls and gave viewers personal and often emotive interpretations of events. Many of these pictures were of major battle scenes and had been painted by artists who had first-hand experience of the conditions the Australians had fought under and of the general atmosphere at the front line. After the war the commissioning program had been extended and on show were a series of portraits of notable men and bronze sculptures. A new art form, the 'picture model', also made its first appearance. Now known as 'dioramas', they were produced by sculptors and painters to represent significant moments in Australia's military history.[21]

Bean and other founders of the Memorial had always considered art to be a central component of the pictorial collection. It had the capacity to unite strong storytelling with powerful symbolism to tell the story of the Australian experience in war. Art also lent prestige to the new institution, and the opening displays of paintings and sculpture attracted much favourable press and public comment.

One of the most important of these works was Lambert's now famous painting, *ANZAC, the landing, 1915*. This depiction of the landing at ANZAC Cove on 25 April had always been envisaged as a centrepiece of the Gallipoli displays. Lambert had commenced the composition in London in mid-1919, almost as soon as he returned from the Middle East. As he worked through his preliminary ideas, he doubtless had in mind the vivid account that Hedley Howe, a veteran of the landing, had given on the Historical Mission's first day on Gallipoli. Howe's story contained all the drama needed for an important painting, and Lambert decided to use it as the narrative framework for the 3.5 metre–wide picture.

ART, PHOTOGRAPHY AND HISTORY

Artist unknown (possibly Fred Leist)
Cover illustration for *Australian War Photographs: a pictorial record from November 1917 to the end of the war*,
London, 1919
AWM V940.40994 A938

Compiled by Hubert Wilkins, this publication capitalised on public interest and promoted the work produced through the Australian official war photographs scheme.

A PICTORIAL RECORD FROM NOVEMBER 1917 TO THE END OF THE WAR·····

AUSTRALIAN WAR PHOTOGRAPHS

PRICE FOUR SHILLINGS

(Following pages)
George LAMBERT (1873–1930)
ANZAC, the landing, 1915,
London and Sydney, 1920–1922
Oil on canvas, 190.5 x 350.5 cm
AWM ART02873

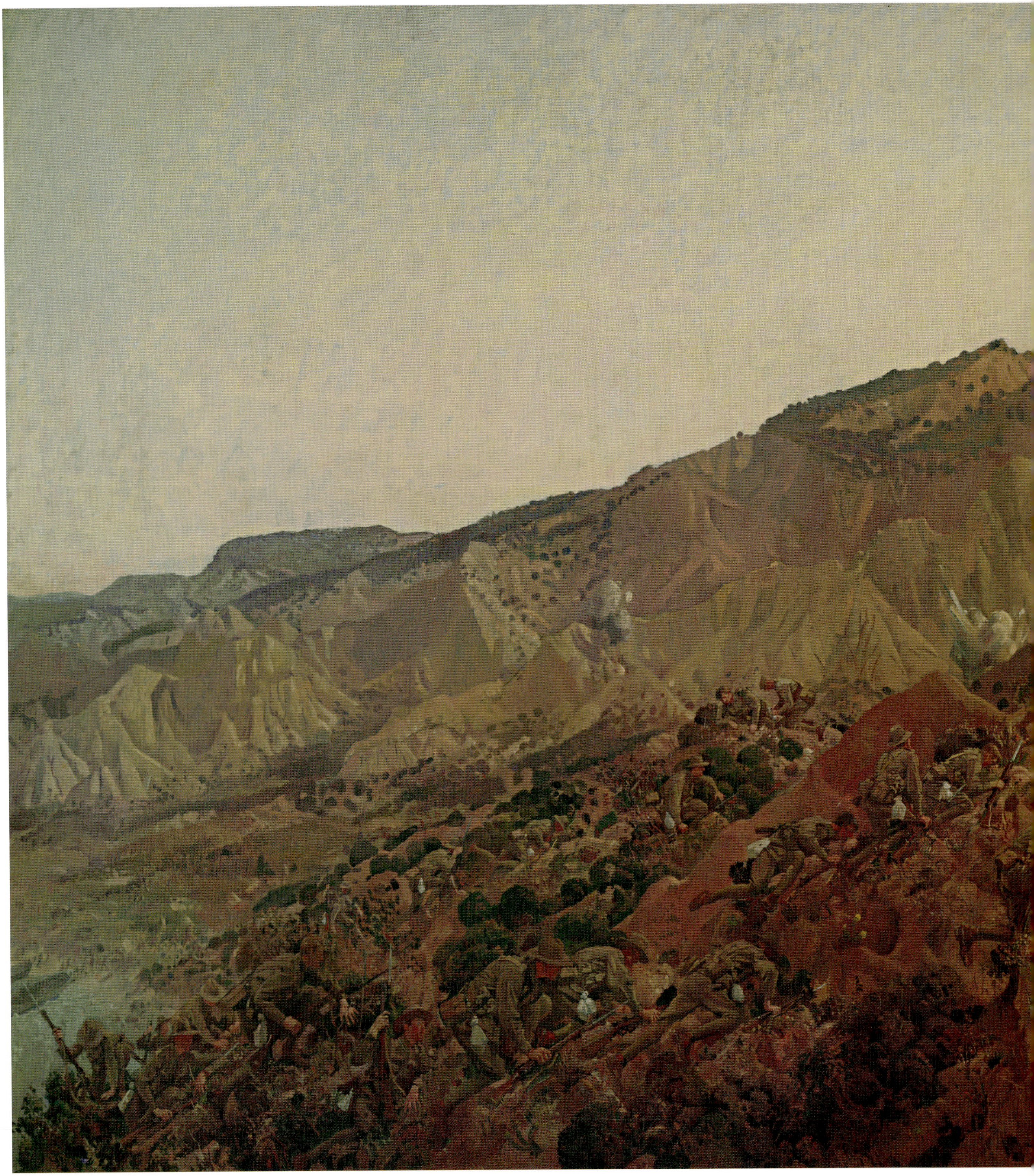

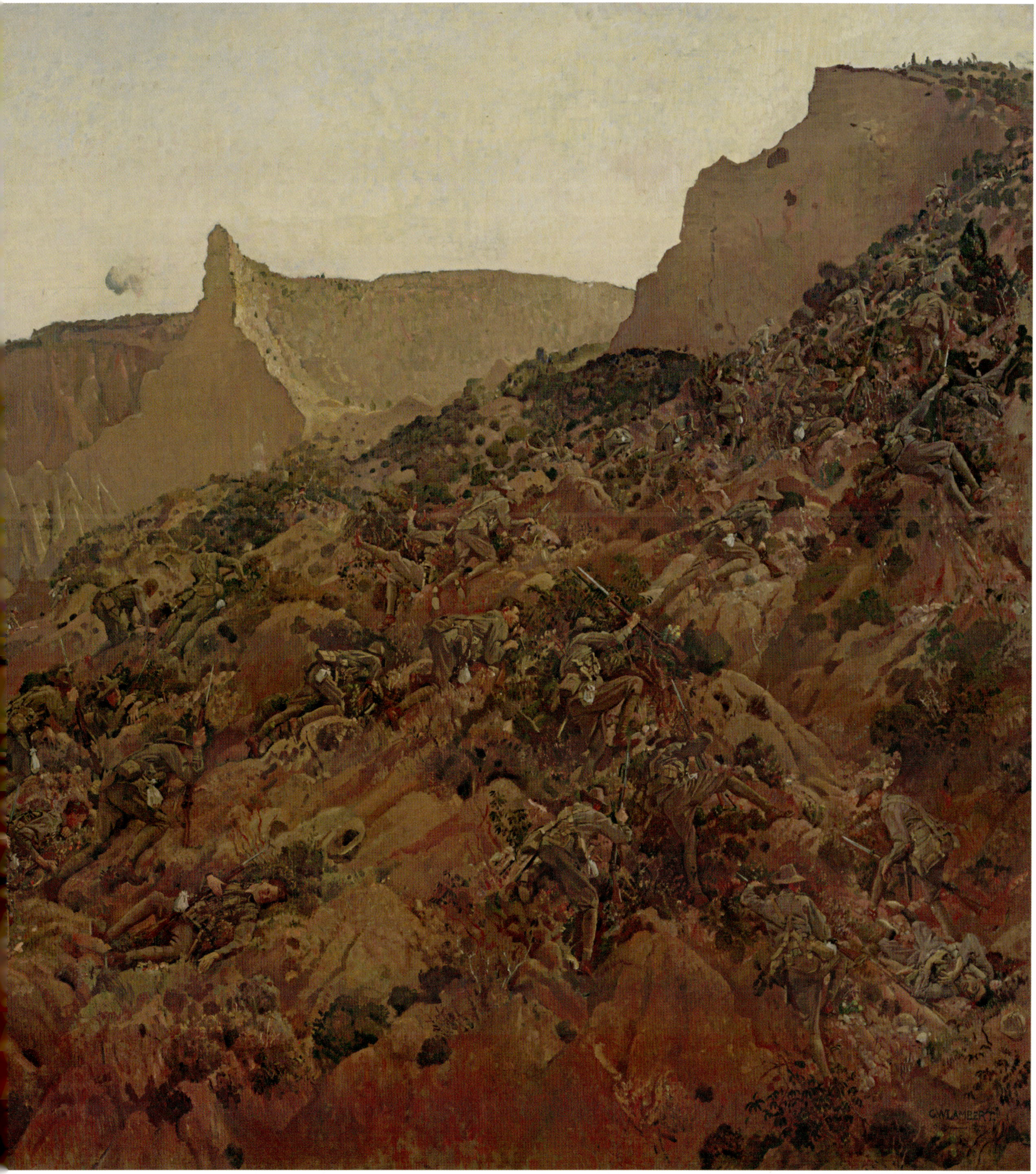

George LAMBERT (1873–1930)
Study for 'ANZAC, the landing, 1915',
London, c. 1920
Pencil on paper, 22.4 x 18.2 cm
AWM ART11391.290

George LAMBERT (1873–1930)
Study for 'ANZAC, the landing, 1915',
London, c. 1920
Pencil on paper, 30.6 x 23.6 cm
AWM ART11391.315

ART, PHOTOGRAPHY AND HISTORY

George LAMBERT (1873–1930)
Study for 'ANZAC, the landing, 1915',
London, c. 1920
Pencil on paper, 21 x 26.6 cm
AWM ART11391.285

A copy of this photograph was used as reference material by Lambert as he worked on the large canvas of the landing.

Hubert WILKINS (1888–1958)
Looking from the crest of Plugge's Plateau, which was first reached on the morning of the landing, across the Razorback to the Sphinx, c. 17 February 1919
Glass whole-plate negative
AWM G01872

With such a broad canvas to work on, Lambert positioned important stages of the story across the painting. Starting in the lower left corner, he showed the Australians landing on the beach in the half light of dawn, the struggle of individuals through the low spiky undergrowth, a young boy dying as others rush past him, and then the scramble up the steeper slopes and the rush across Plugge's Plateau. Lambert fashioned the painting so that the attention of viewers would be drawn back towards the left-hand side by the bursting artillery shell, and then down the rugged ridge lines, right back to the beach in the lower left.

This is an unconventional composition. The strong diagonal line formed by the slope of Ari Burnu divides the picture in two, with the human activity in the foreground distinctly separated from the impressive landscape background. This emphasises the upward rush of troops. The foreground is thrown into shadow, and the use of subdued pinks and khaki colours makes it difficult to see from a distance the many figures as they struggle across the landscape. The artist likened them to a 'small swarm of ants climbing, no matter how rapidly, climbing painfully and laboriously upward'.[22] Lambert also deliberately painted the bodies and the earth in the same tone, making them nearly indistinguishable. From his own experience walking the ground on Gallipoli, he was only too aware that the earth had literally swallowed up many of these men and obliterated their identities.

Lambert used his son Maurice as a model for many of the figures in the composition. Maurice was posed in awkward and almost impossible positions, in an attempt to get the feel of men wounded or dead. After a session modelling as a soldier, he admitted, 'I feel almost as if I had landed at Gallipoli myself'.[23] Like Bean, Lambert confirmed with veterans the specifics of what they had worn on the day, and then borrowed uniforms and kit from stores to get the details right. But strict accuracy was sometimes moderated by artistic concerns in order to create the most powerful picture of the soldiers in action. Whether to show the men in slouch hats or caps troubled Lambert, and he discussed the question at length with Bean and others. As both hats and caps had been worn on the first day, Lambert decided he would show a majority of hats to give a more Australian feel to the painting.

At the time he began work on *ANZAC, the landing, 1915*, Lambert had only a couple of small oil studies that related to the landing area as reference material. He borrowed photographs that Wilkins had taken and used them to render the topographical features of the ridge lines, gullies and other aspects of the landscape. Once again, strict accuracy was sacrificed to achieve the artistic effects Lambert wanted, and he compressed a larger panoramic sweep of the landscape into the painting. The canvas covers an arc of about 240 degrees, much more than the human eye can see in one glance. But his manipulation of the topography allows him to represent the complete story, from the beach where the boats pulled in to the top of Plugge's Plateau. For most viewers, Lambert's artistic licence went unnoticed. The fact that the landscape was distorted and inaccurate was immaterial because it presented an authentic individual story within a landscape that felt right. Incorporated into this painting, Howe's experience became emblematic of the entire Australian endeavour on 25 April and this was subsumed into a greater national narrative.

Lambert returned to Australia in March 1921 with the half-finished painting rolled up and packed in a crate. Memorial staff were anxious for him to complete the commission, and it was finally framed and ready just in time for the opening of the museum. The painting was a key feature of the Gallipoli court. As an early twentieth-century version of traditional nineteenth-century history paintings that focused on tragic and heroic action, Lambert's picture brought together elements of the Gallipoli story in ways that other visual material and objects could not. Its focus on the harsh and difficult terrain rather than on hand-to-hand combat further cemented the idea that the real 'enemy' to be overcome was the landscape.

The public welcomed this painting as one of the first major visual interpretations of a defining moment of the Gallipoli campaign. Although photographs taken by soldiers on Gallipoli had circulated among families and in the press since 1915, these were mostly hurried snaps and could neither represent nor synthesise into a cohesive narrative

George LAMBERT (1873–1930)
Study for 'The charge of the 3rd Light Horse Brigade at the Nek, 7 August 1915', London, c. 1920
Pencil on paper, 25.8 x 36.8 cm
AWM ART11391.308

the events of the landing. Lambert's painting captured the entire sweep of the story and this enormous canvas created a dramatic and emotional experience for veterans and general visitors.

The commission was clearly an outstanding success, the painting being reviewed in the Melbourne *Herald* as a work that conveyed a 'declaration of sacrifice and achievement in a way that no other war picture has done'.[24] The Memorial's sales section also took the opportunity to make a colour print of the painting and this was available, ready for framing, for seven shillings and sixpence. The advertisement described *ANZAC, the landing, 1915* as 'the first authentic picture of this great exploit ... a great work of art and a truthful portrayal of this brilliant achievement'.[25] On 25 April 1924 a photograph of it on display in the museum featured on the front page of *The Sun News-Pictorial*. It showed a museum worker explaining the picture to a young boy with the accompanying caption noting how the 'fine ideal and an imperishable memory' of the deeds of Australian soldiers would be passed on to the youth to be maintained in peace 'or if need be, in war'.[26] The painting was on view in Melbourne for two years, where it was seen by about 750,000 visitors.

In late 1918 Bean had promised Lambert that two Gallipoli paintings would be required for the museum collection. These would be nationally important works and would assume a central place in the collection and in future displays. The second commission was for a large painting showing the Australians charging Turkish positions at the Nek on 7 August 1915. Lambert had made a quick oil sketch of the open ground at the Nek when he visited it with the Historical Mission on 17 February 1919. This was the place where he noted how the 'lines of gallant Australians went down to a man'.[27] He felt that the landscape was a wonderful setting for the tragedy, with the tiny battlefield framed by the gently sloping ground leading up to the Turkish positions and beyond that to Chunuk Bair. In the far distance, a distinctive blue ridge line forms a dramatic backdrop.

Lambert began working on ideas for this painting even while he was travelling with the Historical Mission through Turkey on the way to Egypt. A very sketchy outline on the back of one of his paintings shows his initial ideas for the composition.[28] These were refined in a slightly later drawing but the basic design remained unchanged. *The charge of the 3rd Light Horse Brigade at the Nek, 7 August 1915* employs

Hubert WILKINS (1888–1958)
The Turkish monument, erected on the 'Nek' shortly after the evacuation in honour of those who fell in the severe fighting around this area, c. 17 February 1919
Glass half-plate negative
AWM G01743

(Following pages)
George LAMBERT (1873–1930)
The charge of the 3rd Light Horse Brigade at the Nek, 7 August 1915, Sydney, 1924
Oil on canvas, 152.5 x 305.7 cm
AWM ART07965

Before his departure for Gallipoli, Lambert was asked to paint a large canvas of the tragic incident at the Nek on 7 August 1915, when four successive lines of Australians charged towards the enemy lines and were met with a torrent of gunfire. The fighting was over within an hour; more than 300 Australians died in this brief, savage encounter. In its futility, if not for its scale, this charge was one of the great tragedies of the First World War. The dead were not buried until after the war.

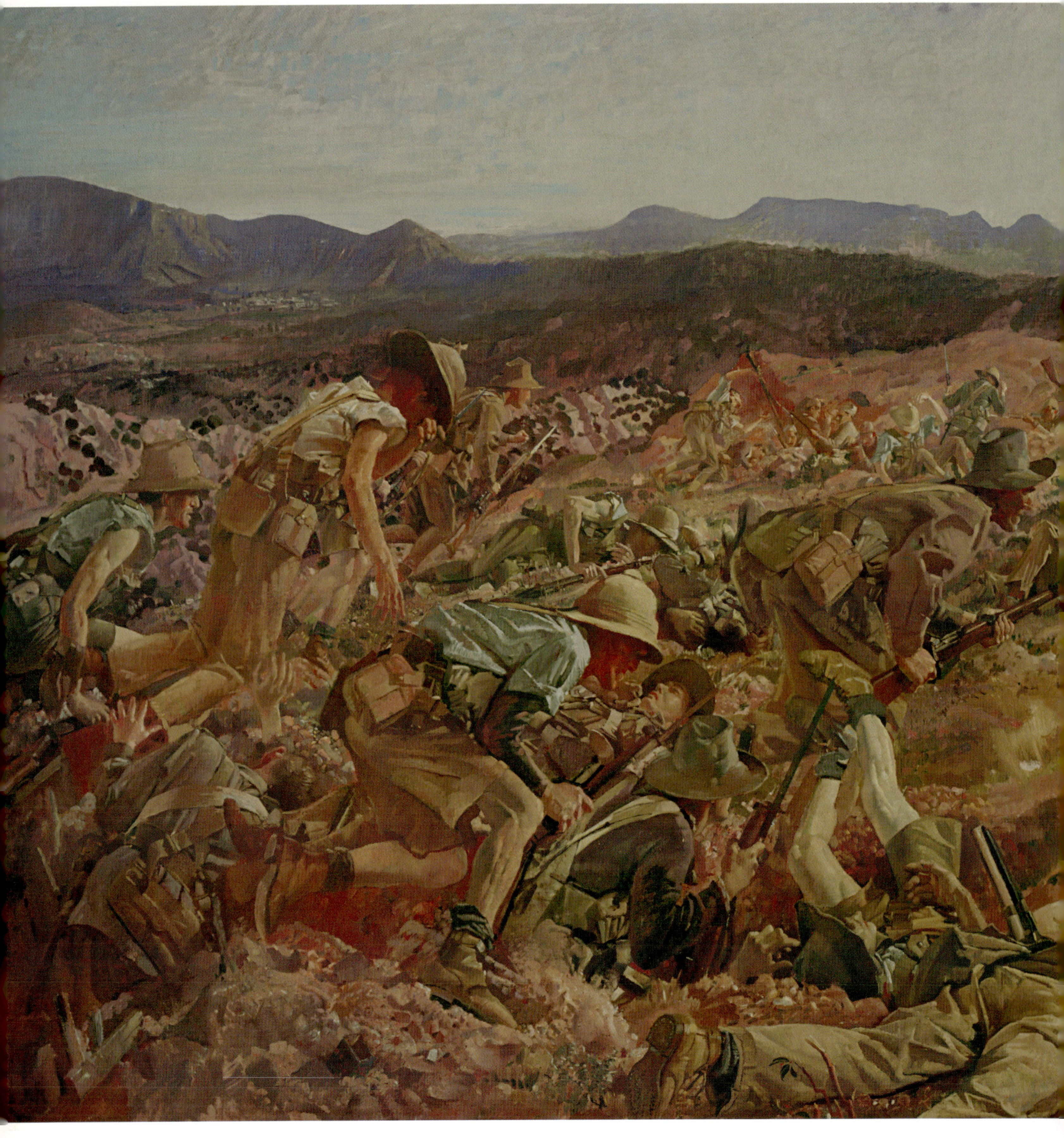

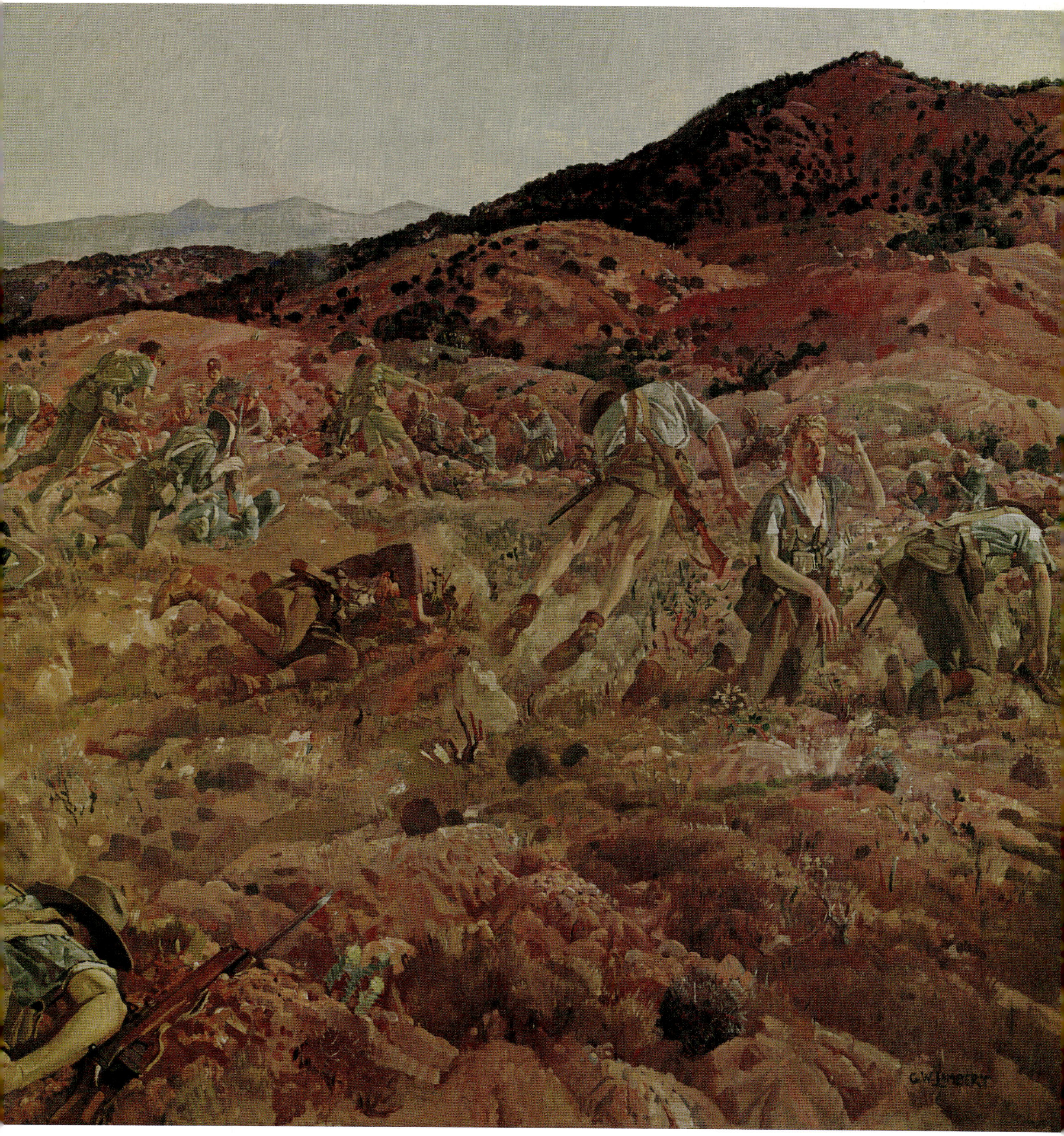
G.W. LAMBERT

George LAMBERT (1873–1930)
Figure study for 'The charge of the 3rd Light Horse Brigade at the Nek, 7 August 1915', London, c. 1920
Pencil on paper, 25.8 x 36.8 cm
AWM ART11391.306

an unusual composition by adopting the viewpoint of a soldier just on the edge of the battlefield. In front are Australians falling, some being spun around by the force of the striking bullet – 'like marionettes jerked into eternity', as a Sydney *Guardian* critic wrote.[29] Others lie dead across the field, while only a metre or so from the Turkish front line men are being shot down. A man on his knees in the centre-right of the picture looks stunned and disbelieving as he raises a wounded hand towards his head. It is a scene of absolute carnage, made more powerfully affecting by the dominant blood-red colours and the churned-up earth.

Lambert wrote to Bean that the picture was 'the biggest job of my life'.[30] As it neared completion, he looked forward to showing it to Bean, for it epitomised 'the Gallipoli Spirit' of the Australian forces.[31] While Lambert was committed to finishing the painting, he was also juggling his work for the Memorial with a number of private commissions. The Memorial had decided to close its displays in Melbourne and reinstall them in Sydney, and throughout 1923 staff again urged Lambert to deliver the painting in time for the opening. A letter from the Art Committee was drafted to stress that as time passed, the reliability of witnesses and indeed of the artist's own memories of the colour and atmosphere of Gallipoli would diminish.[32] To prevent Lambert's taking offence, Bean intervened and a signed contract and schedule for the painting's delivery was quickly obtained from the artist. The finished painting was in place when the Memorial launched its exhibition in Sydney in April 1925.

Like *ANZAC, the landing, 1915*, this second major painting was well received. Lambert was lauded for achieving a dramatic interpretation of the charge at the Nek. Public comment was particularly directed to its sense of 'grim reality' and how the scene was 'stripped of all its glamour'.[33] Bean called it a 'terrible work' for its chilling representation of the event. Much of this can be attributed to Lambert's use of the soldier's viewpoint, which produces a dramatically different effect from that achieved in the earlier work. In *The charge of the 3rd Light Horse Brigade at the Nek, 7 August 1915* the viewer is virtually thrust into the scene and witnesses the slaughter close-up.

Lambert's use of colour had also changed in the couple of years since he had worked on the first painting. Although the painting of the Nek represents a dawn scene and the light is diffused, more white pigment is used and this gives it a chalky, arid feeling. The similarity to the colour of Australian landscapes was noticed by reviewers. This is perhaps not surprising as in the early 1920s Lambert was contemplating a major series of Australian bush pictures.[34] The figures in the painting of the Nek were also thought to be quintessentially Australian, representing 'shearers, boundary riders, rabbit trappers, teamsters' who by their death 'welded six states into a nation in August 1915!'[35]

In the 1920s, many viewers would have agreed with the sentiment expressed by Prime Minister Billy Hughes during the war that Australia had become a nation on the shores of Gallipoli. Lambert's paintings were created at a time when commemoration and nationalism were strong trends in Australian society. For soldiers and families with a personal connection to the events on Gallipoli, paintings provided tangible focal points to remember or visualise what had occurred. In some small measure, paintings could help them begin to reconcile their loss within a specific landscape. The Memorial's displays evoked powerful responses from people. In one instance, a Sydney newspaper reported ex-servicemen 'making a bee-line for the door, tears welling in their eyes' because some painful memory had been ignited.[36] Mothers were said to have found solace in paintings that depicted the places where their sons had last been seen, and newspapers wrote that displays would revive memories or show people 'things unseen before, but spoken of so often as to be almost familiar'.[37]

SIGNIFICANCE OF THE HISTORICAL MISSION

During the interwar years Lambert's paintings were seen by millions of visitors to the Australian War Memorial in its temporary premises in Melbourne and then Sydney (until 1935). Hopes for a permanent memorial and museum building in Canberra were delayed by fluctuating public and government enthusiasm for the project, anti-war sentiment, the Great Depression and the start of the Second World War. In 1941, however, the Memorial finally opened permanent displays in its own building at the foot of Mt Ainslie in Canberra. The original guidebook to the Memorial described how it was 'conceived on the battlefront, born amid the guns at Bullecourt' and 'raised by the living members of the Australian forces to their fallen mates'.[38] In essence, the entire collection was considered to be a memorial to the work of Australian servicemen and women.

The extensive First World War displays that Bean and John Treloar put in place for the Memorial's opening have gradually been reduced to make room for new exhibitions that cover the Second World War and more recent conflicts. However, many items collected by the Historical Mission, or created as a result of it, remain central and iconic features of the Memorial's displays. Lambert's paintings are still feature items in the Gallipoli gallery, and have been since its inauguration. Alongside them now are examples of Wilkins's photographs taken on the peninsula in 1919 and some of the work of James and Swanston. Objects collected by the Historical Mission on Gallipoli are also on view; these include logs retrieved from the Turkish trench coverings at Lone Pine, colour patches, spent bullets, barbed wire and damaged pieces of kit. One of the salvaged lifeboats was shipped back to Australia and is now in the Memorial's Orientation Gallery and seen by nearly a million visitors each year.[39] Gathered from the beaches, trenches, ridges and valleys around the ANZAC area, these items are all part of the evidence collected to help answer some of the lingering questions concerning Australia's part in the Gallipoli campaign. Just as Bean hoped it would, the collection has become a lasting memorial to the tragedy of war as well as to the achievements and sacrifices of the members of the AIF.

The work of the Historical Mission was a small but important part of the greater record-gathering project that Bean and others pursued both during and after the First World War. Even Bean, who used the material created by the

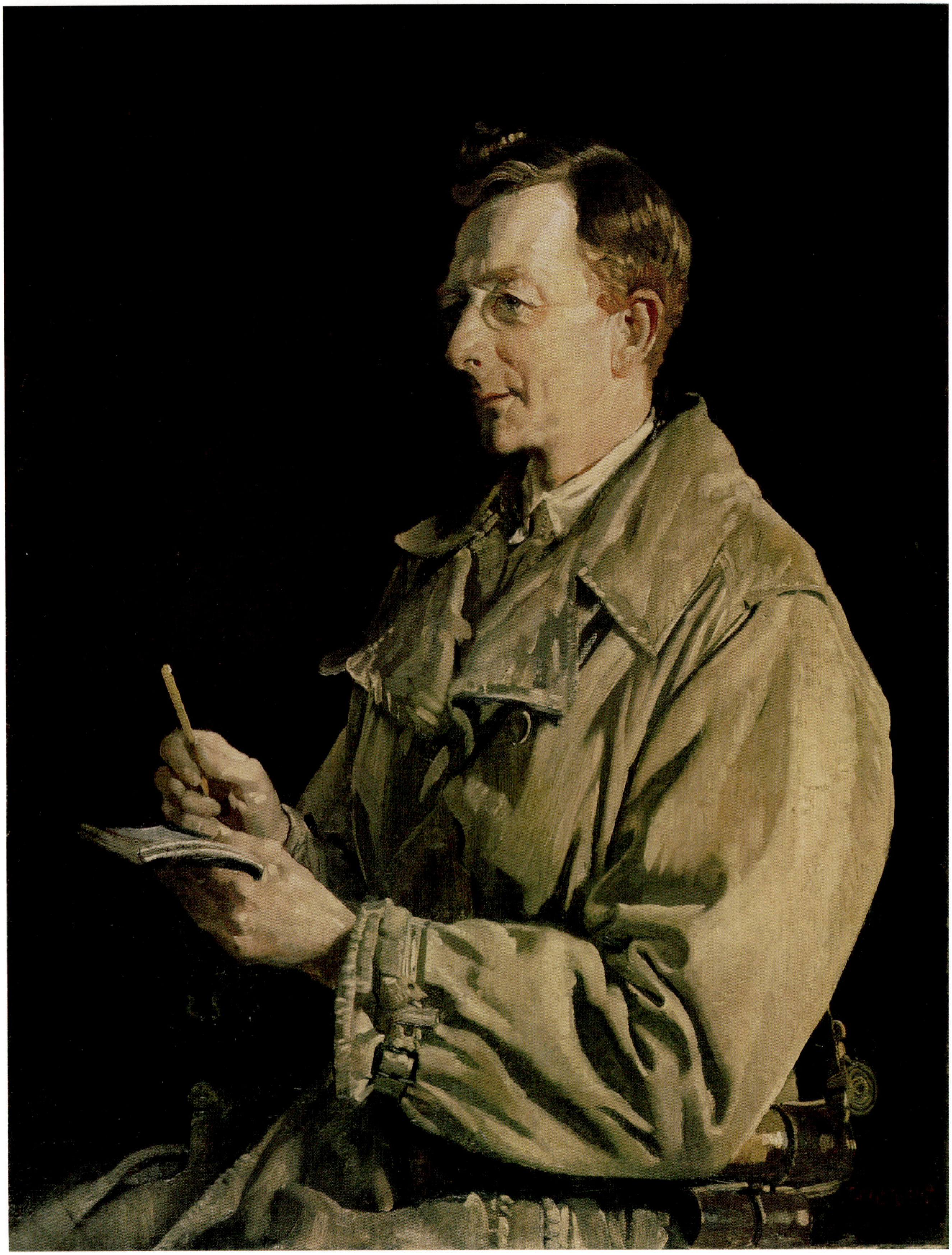

Historical Mission throughout his writings, rarely footnoted it as having its origins in the 1919 expedition. Nevertheless, the Historical Mission created the core collection of objects, records and images about this part of Australia's history, providing the evidence that Bean placed so much emphasis on in his quest for truth and historical authenticity.

Australian attitudes to the Gallipoli campaign and to war in general have changed dramatically since the time of the Australian Historical Mission. Since then, the critical elements of the ANZAC story have constantly been examined and reinterpreted by historians, politicians and cultural commentators. Publications, both popular and scholarly, have analysed the military details; others have discussed the war's impact on Australian society and the legacy of grief and associated public commemorative rituals. In the field of creative arts, Australia's involvement on Gallipoli has been a stimulus for artists, writers and filmmakers, allowing them to re-imagine what it was to be there and to raise issues of national identity. The deaths of the last ANZACs may have severed any direct connection with the event, but public interest in Gallipoli remains strong, with increasing numbers attending the ANZAC Day dawn services both within and beyond Australia.

Australia's involvement in subsequent conflicts such as the Second World War, and then in Korea, Malaya, Indonesia, Vietnam, Iraq and Afghanistan, as well as peacekeeping activities across the globe, has undoubtedly influenced community attitudes to the ANZAC campaign. Yet visitors to the Memorial continue to find relevance and meaning in the Gallipoli displays, now situated within the much larger story of the Australian experience of war.

For nearly a century the rich and varied collection of material assembled for the Australian nation by the Historical Mission has been accessible to public visitors, museum curators, historians, researchers, authors and publishers. These extraordinary paintings, photos, relics and maps have played a vital role in our understanding of this defining moment in Australian history. Like all primary-source material, the collection will always be open to interpretation, and as attitudes to war and the Gallipoli campaign gradually shift, new generations will continue to construct fresh meanings for their own time.

George LAMBERT (1873–1930)
Charles E.W. Bean, Sydney, 1924
Oil on canvas, 90.7 x 71.1 cm
AWM ART07545

In 1924 Lambert commenced a portrait of Charles Bean for the Australian War Memorial. Bean was unwell at the time and postponed some sittings so the artist had to urge his subject to take the task seriously. In response Bean sent a note: 'Dear Lambert, your message received. Does operation start this Saturday, or do we attack on Monday?'

From left to right: Herbert Buchanan, Zeki Bey, William James, Charles Bean and George Lambert.

Hubert WILKINS (1888–1958)
The Australian Historical Mission at lunch on Hill 60,
22 February 1919
Glass 10" x 12" negative
AWM G01904

EPILOGUE

THE AUSTRALIAN HISTORICAL MISSION DISBANDED IN Cairo in April 1919, but what happened to the men after their tour of Gallipoli? This epilogue provides a short summary of the subsequent lives and careers of the AWRS photographers and the men of the Historical Mission.

Charles Bean returned to Australia in 1919 and established himself and the staff who would work on the official history of the war at Tuggeranong, near Canberra. Working life at Tuggeranong was interspersed with occasional social and sporting occasions and in early 1920 he met Ethel 'Effie' Young at a tennis match. They married on 24 January 1921 at St Andrew's Cathedral, Sydney. The same year, the first volume of the official history was published, with the second volume following three years later. The series took 23 years to complete; in all, Bean wrote half of the 12 volumes and coordinated and edited the others. He also worked on several volumes of the related medical services history of the war.

Bean's interest in the institution he helped found, the Australian War Memorial, never wavered. He was always there, serving on committees, providing advice and constantly encouraging others to add original material to the collections. He was also there on 11 November 1941, when the Memorial's new building was opened in Canberra. Australia's involvement in the Second World War prompted him to produce several small publications, including *The Old AIF and the New* (1940) and *War Aims of a Plain Australian* (1943).

Even after the completion of the official history series in 1942, Bean continued to provide accessible and personal accounts of the war, and in 1946 he wrote a condensed single-volume history of the First World War, *ANZAC to Amiens*. In 1948 he returned to the subject of Gallipoli with *Gallipoli Mission*, his book about the Australian Historical Mission. He also became involved in a number of public organisations, helped found the Parks and Playgrounds Movement of NSW and the Commonwealth Archives, and between 1947 and 1958 he chaired the Promotion Appeals Board of the Australian Broadcasting Commission. While on a visit to Britain in 1951, Bean was offered the Chairmanship of the Board of Management of the Australian War Memorial, an unpaid position he held until September 1959. In 1964 he was admitted to Concord Repatriation Hospital, where he died in August 1968.

Arthur Bazley MSM continued to work as Bean's 'deviller', keeping the records in order and chasing up extra bits of information for the historian until 1939. Throughout his life Bazley wrote regularly for the ex-servicemen's magazine *Reveille*, providing thorough, accurate and interesting biographical notes or articles on Australian battles. From 1950 to 1967 he was editor for the RSL magazine *Stand To*.

As with many of the people associated with Bean and the AWRS, his passion for collecting, ordering, and making available the documents that could help Australians understand the First World War continued well after 1918. It was a natural progression that some of the AWRS staff would take a leading role in the management of the collection they had helped form. In 1940 Bazley was appointed chief clerk and librarian at the Memorial. Bazley was Acting Director from 1942 to 1946. John Treloar, who had been appointed Director of the Memorial in 1920, took leave during the Second World War to command the Military History and Information Section (later Military History Section) for the army. Bazley was involved in the appointment of official artists and photographers, and organised exhibitions to present the new work to the public as soon as it was available.

In public life he was involved in various other organisations, including the Commonwealth Archives. Shortly after joining the Department of Immigration in 1946, he became Secretary of the Commonwealth Immigration Advisory Council. In 1949, eight years after the death of his first wife, Annie, he married Mary McPhee. He retired in 1961 and died in Canberra Hospital on 31 July 1972.

John Balfour MBE (Military), described by Bean as 'compact in body and mind', went on to work with the historian until the completion of the official history series in 1942. Bean paid tribute to Balfour's 'great "back-stopping" feat in checking by every possible means each reference among the many thousands in the volumes, all of which he indexed'.[1] From 1942 to 1944 Balfour worked at the Australian War Memorial and then transferred to the army's Military History Section, again working under Treloar. During the Second World War he worked mainly in the South-West Pacific Area and the Philippines, collecting documents and identifying objects for the Memorial's collection. He was present at the signing of the Instrument of Surrender on USS *Missouri* in Tokyo Bay on 2 September 1945 and took a series of photographs documenting the historic event.

In June 1946 he returned to Australia and soon took up a position as a senior research officer for the official historian of the Second World War, Gavin Long. His patience, experience in records work and capacity 'to select the "meat" and resolve any problems as they crop up' was invaluable to the team. After spending 11 years working on the histories, Balfour retired and led a quiet family life in Canberra. He died in Canberra in 1976.

In the years following the First World War, **Herbert Buchanan** moved around New South Wales and Victoria. Most likely these changes of location were related to his profession as a civil engineer. Between 1920 and 1950 he lived successively in Wangaratta (NSW), St Kilda (Vic.), Lindfield (NSW), Newcastle (NSW) and Glen Iris (Vic.). When publishing the *Gallipoli Mission* book in 1948 Charles Bean noted: 'H.S. Buchanan, after alternating private with government practice, is Deputy Chairman of the Federal Contract Board, and has lately given some thrills to his old comrades, as a member of the national radio quiz team.'[2] Buchanan died in Melbourne in 1967, aged 77.

George Hunter Rogers returned to Australia and worked for Bean until the end of 1919, preparing the maps and sketches required for publication in the official history. He resumed civilian life as an engineer with the Public Works Department of Victoria. Towards the end of the 1920s, he became a consultant for private construction companies and even worked at designing and manufacturing small appliances. Somewhere around 1934 he became interested in aerial surveying and was one of the first in this field in Australia. During the Second World War Rogers was employed by the Harbour Trust (Melbourne), managing the construction of slipways and wharves. In his retirement he lived in the Mornington Peninsula area and wrote *The Early History of the Mornington Peninsula* (1960) and a *History of the 'Woodlands' Golf Club* (1984). He self-published an autobiography titled *My First Eighty Years* in 1982 and died three years later.

Hedley Vicars Howe, 'the young scallywag' who had taken part in the ANZAC landing and then guided the Historical Mission through his part of the day's events, returned to Western Australia on 27 April 1919. He went back to working in the pearling industry in Broome, and over the next couple of years kept in contact with Charles Bean. In 1926 Howe left Broome and headed for New Guinea but was persuaded to stop over in Sydney. He was offered a job as private secretary to ex-prime minister Billy Hughes and remained with him for four and a half years. In 1930 Howe married Marjorie Hope and took up a position with the New South Wales Chamber of Manufacturers. Resuming war work in 1940, Howe became a staff captain (and later a major) in Sydney training depots and in 1941 was appointed Military Secretary to the Minister for the Army. He resumed his civilian career with the Chamber of Manufacturers in 1946, before retiring in 1953 and purchasing an orchard at Dural, close to Sydney. He died in 1977.

For **George Lambert**, the official artist of the Historical Mission, the journey to Gallipoli had been a prelude to work that would take several years to complete. In August 1919 he travelled from Cairo to London, where he commenced the major official painting commissions. He made preliminary sketches and a scale drawing of the final composition for *ANZAC, the landing, 1915*, which was transferred to the prepared canvas. In 1921 Lambert returned to Australia to work on the paintings. The art community welcomed him back, and through the 1920s numerous articles and exhibitions celebrated his work.

The completed painting of the Australians landing at Gallipoli was included in the inaugural Memorial displays in Melbourne in 1922. Other battle pictures for the Australian War Memorial followed, including *The charge of the 3rd Light Horse Brigade at the Nek, 7 August 1915*, which *was* delivered in 1924. The Memorial remained a strong patron of his work and in the same year commissioned Lambert to paint a portrait of Charles Bean, a job that Lambert described as carrying 'much joy & excitement in the prospect'.[3] He organised several sittings with Bean, and when Wilkins visited New South Wales the three friends were able to meet up.

While working on the Memorial's commissions, Lambert painted landscapes and portraits for private clients and in 1927 won the Archibald Prize with his painting of Mrs Annie Murdoch. He became interested in sculpture and completed memorials for Geelong Grammar School and St Mary's Church, Sydney, and a memorial to Henry Lawson still on display in the Royal Botanic Gardens, Sydney. A sculpture of Australian Light Horse soldiers that Lambert started in 1919 for a competition was purchased by the Memorial, and through the late 1920s Lambert worked on preparing it for casting in bronze.

Horses remained an important part of his life and it was while tending to his horse's feed-box at Cobbity, New South Wales, on 29 May 1930 that he had a heart attack and died. Two memorial exhibitions were held in 1930, but his work gradually lost favour with many art critics and collectors. However, in 2007 it became known to a new audience through two major exhibitions in Canberra. The National Gallery of Australia presented a thorough retrospective of his work and the Australian War Memorial mounted an exhibition of his Gallipoli and Palestine landscapes from the 1918–19 period. Together the exhibitions presented a cohesive overview of Lambert's work, which gave the Australian public an opportunity to reassess his contribution to Australian art.

In 1919, after completing his Gallipoli work, **William Hopkin James** returned to Britain and took a short course in Wales in hardwood timber milling as preparation for his civilian life. On his return to Australia he became a sawmiller and in 1940 he listed his occupation as sawmill proprietor in Mayfield, near Newcastle, New South Wales. He stayed in the Newcastle area and in 1965 was living in Adamstown. James had first enlisted in the Light Horse in 1905, and during the interwar period his military service continued with the Citizen Military Forces. He commanded the 16th Light Horse and then the 16th Machine Gun Regiment. In 1938 he was nominated to lead a hand-picked group of 80 light horsemen called the Centenary Regiment to parade in the sesquicentenary celebrations of New South Wales.

Aged 52 when the Second World War broke out, he reported for war service at Newcastle. He attended Senior Officers' School and volunteered for active service in 1941 and transferred to the AIF. In 1943, however, he was placed on the retired list of the Light Horse Reserve, Citizen Military Forces, with the honorary rank of colonel. This ended 38 years' continuous service with Australian military units. William James MC died in New South Wales, 29 December 1972. His son, Brigadier J.A. James, carried on the family tradition of military service and was appointed first Brigadier of the Royal Australian Armoured Corps in 1957.

Until recently, hardly anything was known about the AWRS photographer **William Henderson Swanston** and his background as a professional photographer. On his return from Gallipoli the AWRS extended his secondment so that he could complete the printing of photographs for William James. He left Egypt for Britain on 26 May 1919 and in July transferred to the 'G' Reserve of the RAF. On 30 April 1920 he was discharged from the air force. It is very likely that he returned to Edinburgh and continued the family photographic business after his father died in May 1921. A trade directory indicates that the family-run business of Swanston Photographics continued until 1930.

After working on the Historical Mission, **Hubert Wilkins** returned to the Western Front to complete his photographic survey of battlefields relating to Australia's military actions. Bean had once said of Wilkins: 'It is easily safe to say that his is the most adventurous life that has been lived in our generation.'[4] His work during the war was clearly dangerous, but his postwar adventuring earned him international respect and a number of high-level awards, including a knighthood. In late 1919 he entered the England to Australia air race with its £10,000 prize. Unfortunately, his plane 'Kangaroo' crashed in Crete in November 1919. Wilkins was unhurt – and undeterred. Keen to push the limits of aviation, he set out to explore the polar regions by plane, embarking on several schemes in the early 1920s to explore both the Antarctic and the Arctic. The British Museum sponsored Wilkins on a natural history expedition to northern Australia from 1923 to 1925, during which he collected thousands of specimens. His book *Undiscovered Australia* was published in 1928. Partnering with Carl Ben Eielson, Wilkins realised his dream of flying across the Arctic in 1927; then a year later they were the first to fly over Antarctica. In 1929 he married Suzanne Bennett, an Australian-born actress who lived in New York.

Wilkins made many pioneering flights in aircraft and zeppelins before turning in the early 1930s to using submarines for his Arctic expeditions. A passion for geography and the study of weather led to a contract with the US Weather Bureau and the Arctic Institute of North America during the Second World War to investigate the possibility of establishing meteorological bases in polar areas. Somewhat bizarrely, he was also interested in telepathy and conducted several experiments in long-range telepathy, which he discussed in his book *Thoughts through Space* (1951). Sir Hubert Wilkins MC and Bar died on 30 November 1958 in Massachusetts, USA. As a mark of respect, his ashes were scattered at the North Pole by the crew of a US nuclear submarine.

NOTES

INTRODUCTION

1. C.E.W. Bean, 'The beginnings of the Australian War Memorial', n.d., AWM38: 3DRL 6673/619.
2. ibid.
3. ibid.
4. Honorary Captain Charles Edwin Woodrow Bean (1879–1968), born in Bathurst, New South Wales, and travelled with his family to Britain in 1889; attended Clifton College and during his time there joined the School Engineer Corps; studied Classics at Oxford and admitted to Bachelor of Arts (2nd class honours, 1902), Bachelor of Civil Law (3rd class honours, 1904); returned to Australia in 1904; appointed Australia's official war correspondent in 1914; Mentioned in Despatches for actions at Krithia, Gallipoli, in May 1915; appointment with AIF terminated in June 1919. Throughout the book, short biographical notes for all the principal military personnel mentioned are included at the first mention of the name. These notes have mainly been taken from military service records and indicate service in the First World War only. Rank and any other decorations are given as they were at time of discharge from the service.
5. *Literae Humaniores*, or 'Greats', as it was commonly called.
6. C.E.W. Bean, 'The beginnings of the Australian War Memorial', n.d., AWM38: 3DRL 6673/619.
7. Honorary Captain George Washington Lambert (1873–1930) enlisted 13 December 1917 in London AIF Administrative Headquarters Unit as official artist; worked in Middle East and on Gallipoli; having completed his work, his appointment as official artist was terminated on 31 March 1920.

    Captain George Hubert Wilkins MC and Bar (1888–1958) enlisted 1 May 1917, Australian Flying Corps (AFC), 9th Reinforcements; embarked Sydney on HMAT *Marathon* A74; transferred to AWRS London as official photographer on 21 August 1917; worked on the Western Front until appointment as official photographer was terminated on 7 September 1920 having completed duty; prior to First World War was official correspondent to Balkan War of 1912–13 and member of Vilhjalmur Stefansson's Arctic exploration party (1913–16).
8. Lieutenant William Hopkin James MC (1887–1972) enlisted 22 August 1914, 1st Australian Light Horse, service no. 81; embarked Sydney on HMAT *Star of Victoria* A16; served on Gallipoli 12 May to 19 December 1915, then in Middle East; transferred to the AWRS Cairo on 6 November 1918; discharged from service on 10 February 1920.

    Sergeant William Henderson Swanston (1881–unknown) enlisted 7 January 1916 with No. 14 Squadron, Royal Flying Corps, service no. 17582; served in Middle East; later seconded to AWRS Cairo c. 27 November 1918 to April 1919; discharged from service on 30 April 1920.
9. Bean's own account was published in *Gallipoli Mission*, Australian War Memorial, Canberra, 1948.

CHAPTER 1

1. *Gallipoli Mission* was first published by the Australian War Memorial, Canberra, in 1948, reprinted in 1952 and again in association with ABC Enterprises in 1990.
2. C.E.W. Bean, *Gallipoli Mission*, p. 1 [all subsequent references are to the first edition].
3. C.E.W. Bean, *Gallipoli Mission*, p. v.
4. Sergeant Humphrey Courtenay Gilbert Kempe (1895–1987) enlisted 24 October 1914, 3rd Australian Light Horse Regiment, 1st Reinforcements (later 1st LH Bde MG Sqn), service no. 627; embarked Newcastle on HMAT *Karoola* A36; served on Gallipoli c. 25 April to c. 1 December 1915, then in the Middle East; discharged from service on 23 February 1919.
5. H.G.C. Kempe, letter to C.E.W. Bean, 25 July 1948, 'Pfalz incident', AWM38: 3DRL 6673/203.
6. ibid.
7. A.W. Bazley, 'Australia's Official History of World War 1', *Stand To*, vol. 6, no. 6, November 1958 to January 1959, p. 23.
8. 'Message from the King. Congratulations on bravery', *The Argus*, 1 May 1915, p. 17.
9. C.E.W. Bean, 'Gallipoli. How the Australians fought. Imperishable fame', *Sydney Morning Herald*, 15 May 1915, p. 14.
10. C.E.W. Bean, 'Gallipoli. On the beach. The morning duel', *Sydney Morning Herald*, 17 July 1915, p. 14.
11. C.E.W. Bean, diary entry, 27 November 1914, AWM38: 3DRL 606/1, p. 97
12. C.E.W. Bean, letter to Commander Pethbridge, 16 October 1914, AWM38: 3DRL 6673, Item 270.
13. Charles Bean spent short rest periods at the Press Correspondents' camp on nearby Imbros from July onwards and also travelled to Mytilene (Mitilíni), the main harbour of the Greek island of Lesbos, in September 1915.
14. C.E.W. Bean, diary entry, 20 December 1915, AWM38: 3DRL 606/22.
15. ibid.
16. Australian deaths at Gallipoli were 8,709, with a total of 61,522 Australian deaths overall in the First World War.
17. Senator G. Pearce, Australian Minister for Defence, 'What ANZAC means', *The Argus*, 25 April 1916, p. 5.
18. ibid.

19. Prime Minister W.M. Hughes, 'On the heroes of Anzac' in '*The day' – and after: war speeches of the Rt. Hon. W.M. Hughes*, Cassell and Company Ltd, London, 1916, p. 68. 'Boldly they rode and well,/Into the jaws of Death,/Into the mouth of Hell/Rode the six hundred' are some of the best known lines from Tennyson's poem.
20. Prime Minister W.M. Hughes, addressing Australian troops in France, 2 July 1918, AWM38: 606/116, p. 12.
21. Signaller Ellis Luciano Silas (1884–1972) enlisted 28 September 1914, 16th Battalion, service no. 634; embarked Melbourne on HMAT *Ceramic* A40; served on Gallipoli 25 April to 28 May 1915; discharged as permanently unfit for active service on 17 August 1916.

    Lieutenant George Courtney Benson (1886–1960) enlisted 8 September 1914, 3rd Field Artillery Brigade, service no. 2180; embarked Melbourne on HMAT *Geelong* A2 and served on Gallipoli from c. 2 May 1915, and in the Middle East and France; from 14 March 1918, General List AIF as official artist, attached to 4th Division, France.

    Horace Millichamp Moore-Jones (c. 1868–1922) enlisted 1914, British Section of New Zealand Expeditionary Force, 1st Field Company of Engineers; served on Gallipoli from 25 April 1915; declared unfit for war service in 1916 and repatriated to New Zealand.
22. C.E.W. Bean, 'The beginnings of the Australian War Memorial', AWM38: 3DRL 6673/619.
23. C.E.W. Bean, letter to Minister for Defence, Senator G. Pearce, 8 November 1916, AWM93: 12/12/1.
24. Major John Linton Treloar (1894–1952) enlisted 27 August 1914, service no. 5 (later VX39804); served on Gallipoli with 1st AIF Division Head Quarters 25 April to 4 September 1915, later in No. 1 Squadron, AFC, and AWRS from May 1917; discharged from service as major on 17 July 1919; later Director of the Australian War Memorial (1920–1952) and officer in charge of the Military History Section during the Second World War.
25. C.E.W. Bean, 'Australia's records. Preserved as sacred things. Pictures, relics, and writings', *ANZAC Bulletin*, no. 40, 10 October 1917, p. 15.
26. ibid., p. 14.
27. From November 1916 to June 1917 Bean had access to a British press photographer, Captain Herbert F. Baldwin (1880–1920), who became Australia's first official war photographer.
28. Honorary Captain James Francis Hurley (1885–1962) enlisted 17 August 1917 in London, AIF Administrative Headquarters Unit as official photographer; worked on Western Front and Middle East, resigning from AIF on 11 July 1918.
29. Hubert Wilkins, in Lowell Thomas, *Sir Hubert Wilkins: His world of adventure: An autobiography*, Readers Book Club in Association with the Companion Book Club, London, 1961–62, p. 86.
30. C.E.W. Bean, 'Australia's records. Preserved as sacred things. Pictures, relics, and writings', *ANZAC Bulletin*, no. 40, 10 October 1917, p. 14.
31. Honorary Lieutenant William Henry Dyson (1880–1938) enlisted 10 December 1916, 1st ANZAC Headquarters, as official artist; former political cartoonist with the *Bulletin*; later moved to London then recorded AIF activities on the Western Front; discharged on 31 March 1920 having completed duty as a war artist.
32. Dr W. T. Ellis, 'Advance Australia! Wonderful story of valour written on the battlefields of France and Egypt by the Anzacs', *New York Herald*, 10 August 1919, p. 1.
33. C.E.W. Bean, diary entry, 18 January 1919, AWM38: 3DRL 606/229/1.
34. Printed circular, 'War Museum', 3 April 1918, Commonwealth of Australia, Department of Defence, Melbourne, AWM38: 3 DRL 6673/621.

CHAPTER 2

1. C.E.W. Bean, letter to his parents, 12 January 1919, AWM38: 3DRL 7447/7.
2. Bill Gammage, Foreword, *Gallipoli Mission*, ABC Enterprises in association with the Australian War Memorial, Sydney, 1990, p. vi.
3. For a discussion of this see Shaune Lakin, *Contact: Photographs from the Australian War Memorial collection*, Australian War Memorial, Canberra, 2006, pp. 69–71.
4. C.E.W. Bean, letter to his parents, 19 January 1919, AWM38 3DRL 7447/7.
5. ibid.
6. ibid.
7. See C.E.W. Bean, *Gallipoli Mission*, Australian War Memorial, Canberra, 1948, p. 18; and Simon Nasht, *The Last Explorer: Hubert Wilkins: Australia's unknown hero*, Hodder Australia, Sydney, 2005, p. 24–31.
8. The appointment of Hubert Wilkins as a second photographer was suggested by Brigadier General C. T. Griffiths of AIF Administrative Headquarters. C.E.W. Bean, diary entry, 6 August 1917, AWM38: 606/84, p. 48.
9. Lowell Thomas, *Sir Hubert Wilkins: His world of adventure: An autobiography*, p. 87.
10. This story is recorded in John Grierson, *Sir Hubert Wilkins: Enigma of exploration*, Robert Hale Ltd, London, 1960, p. 9.
11. C.E.W. Bean, diary entry, 19 April 1918, AWM38: 3DRL 606/107, pp. 51–53.
12. C.E.W. Bean, 'Australia's records. Preserved as sacred things. Pictures, relics, and writings', *ANZAC Bulletin*, no. 40, 10 October 1917, p. 14.
13. G.H. Wilkins, letter to C.E.W. Bean, 20 November 1923, AWM38: 3DRL 6673/468.
14. John Grierson, *Sir Hubert Wilkins: Enigma of exploration*; Lowell Thomas, *Sir Hubert Wilkins: His world of adventure; An autobiography*; Simon Nasht, *The Last Explorer: Hubert Wilkins: Australia's unknown hero*.
15. G.H. Wilkins, the papers of Sir George Hubert Wilkins, Byrd Polar Research Centre, Ohio State University.
16. A substantial collection of Lambert family papers, including his war diary for 1918–19, is in the Mitchell Library, Sydney, ML MSS 97/4-14X.
17. G.W. Lambert, application to enlist in the British Army, 3 December 1915 in 'Papers re war service 1915–1919', Mitchell Library, Sydney, ML MSS 97/4. No specific reason was given for rejecting his application.
18. G.W. Lambert, letter to Amy Lambert, 15 January 1918, 'Lambert family papers', Mitchell Library, Sydney, ML MSS 97/10.
19. G.W. Lambert, quoted in William Moore, *The Story of Australian Art: From the earliest known art of the continent to the art of

to-day, Angus & Robertson Ltd, Sydney, 1934, vol. 2, p. 62.
20. C.E.W. Bean, diary entry, 20 November 1918, AWM38: 3DRL606/117, p. 83.
21. C.E.W. Bean, *Gallipoli Mission*, p. 16.
22. Staff Sergeant Arthur William Bazley MSM (1896–1972) enlisted 5 October 1914, 1st Division Headquarters, service no. 69; embarked Melbourne on HMAT *Orvieto* A3; served as assistant to Charles Bean on Gallipoli and Western Front, then transferred to AWRS London in June 1917; discharged on 10 August 1919.
23. Captain John Balfour (1892–1976), Corps of Military Staff Clerks 1910–1914; re-enlisted in the same capacity on 16 August 1914, service no. 4; embarked Melbourne on HMAT *Orvieto* A3; served on Gallipoli from 25 April to 3 November 1915; transferred to the 1st Australian Headquarters in the Middle East, then Western Front and, in 1918, AWRS, London; discharged from service on 18 July 1923.
24. Lieutenant Hedley Vicars Howe (1892–1977) enlisted 28 August 1914, 11th Battalion (later Imperial Camel Corps, 1st ANZAC Battalion), service no. 188; embarked Fremantle on HMAT *Ascanius* A11; served on Gallipoli from 25 April to early December 1915, then Middle East and Western Front; transferred to AWRS as part of the Australian Historical Mission on 5 January 1919; discharged from service on 4 July 1919.
25. Lieutenant Herbert Sidney Buchanan (1889–1967) enlisted 30 August 1915, 5th Field Company Engineers (later 14th Field Company, 4th Division Engineers, 1st ANZAC Headquarters), service no. 2969; embarked Melbourne on HMAT *Ceramic* A40; served in Middle East and Western Front, and awarded Belgian Croix de Guerre; discharged from service on 11 June 1919.

    Sergeant George Hunter Rogers (1896–1985) enlisted 17 March 1915 in 6th Field Ambulance (later 12th Field Ambulance), service no. 3357; embarked Melbourne on HMAT *Ajana* A31, served on Gallipoli 30 August to c. 19 December 1915, then Middle East and Western Front; transferred to AWRS, London, Topographical Section, as part of the Australian Historical Mission on 11 January 1919; discharged from service on 10 August 1919.
26. C.E.W. Bean, *Gallipoli Mission*, p. 9.
27. ibid., p. 10.

CHAPTER 3

1. C.E.W. Bean, diary entry, 3 December 1918, AWM38: 3DRL 606/117, p. 85.
2. C.E.W. Bean, letter to his parents, 16 December 1918, AWM38: 3DRL 7447/7-8.
    Honorary Lieutenant Albert Henry Fullwood (1863–1930), volunteer Royal Army Medical Corps, 3rd London General Hospital, 24 April 1915 to 16 November 1917; enlisted c. April 1918, General List AIF, as official artist; attached to 5th Division; worked on Western Front 2 May to 8 August 1918; discharged 31 December 1919, having completed duty as a war artist.

    Staff Sergeant William Henry Joyce MSM (1890–unknown) enlisted 8 October 1916, 2nd Divisional Ammunition Column, 13th Reinforcements, service no. 32091; embarked Sydney on RMS *Osterley*; served in France; transferred to AWRS on 28 June 1918, as official photographer; attached to 4th Squadron, AFC, 3 January 1919; discharged from service 18 December 1919.
3. G.W. Lambert, letter to H.C. Smart, 11 December 1918, AWM93: 18/7/7 part 1. Honorary Captain Henry Casimir Smart attached to AIF Administrative Headquarters as Officer in Charge of Military Records; later Publicity Director in London for Australian government and editor of *ANZAC Bulletin*.
4. C.E.W. Bean, letter to H.C. Smart, 16 December 1918, AWM93: 18/7/7 part 1.
5. ibid.
6. G. Hunter Rogers, *My First Eighty Years*, self-published, Melbourne, c. 1982, p. 62.
7. C.E.W. Bean, *Gallipoli Mission*, Australian War Memorial, Canberra, 1948, pp. 15–16.
8. J. Grierson, *Sir Hubert Wilkins: Enigma of exploration*, p. 212.
9. C.E.W. Bean, *Gallipoli Mission*, p. 17.
10. C.E.W. Bean, letter to J. Treloar, 4 December 1918, AWM16: 4359/1/13.
11. C.E.W. Bean, *Gallipoli Mission*, p. 14.
12. C.E.W. Bean, diary entry, 12 November 1918, AWM38: 3DRL 606/117, p. 80.
13. ibid., p. 78.
14. G. Hunter Rogers, *My First Eighty Years*, p. 62.
15. The Paris Peace Conference opened on 18 January 1919.
16. G.W. Lambert, letter to Mr Box, Secretary to the High Commissioner for Australia, 20 January 1919, 'Papers re war service 1915–1919', Mitchell Library, Sydney, ML MSS 97/4.
17. Captain Henry Somer Gullett (1878–1940) appointed official press correspondent to Western Front 1915; later enlisted 16 May 1916, 2nd Division Ammunition Column Field Artillery, 13th Reinforcements, service no. 32074; embarked Sydney on RMS *Osterley*; appointed head of AWRS, Cairo, in 1917, then official press correspondent to AIF in Middle East; discharged from service on 24 September 1919.
18. J. Treloar, letter to H.W. Dinning, 6 March 1919, 'AWRS EEF Subsection Cairo reports', AWM25: 1013/1 part 5. This commission for Wilkins did not eventuate.
19. C.E.W. Bean, *Gallipoli Mission*, p. 26; C.E.W. Bean, diary entry, 22 January 1919, AWM38: 3DRL 606/229/1.
20. Taranto was an important Royal Navy communications centre and transit base for allied troops during the war.
21. C.E.W. Bean, *Gallipoli Mission*, p. 28.
22. C.E.W. Bean, diary entry, 27 January 1919, AWM38: 3DRL 606/229/1, p. 18.
23. G.W. Lambert, journal entry, 3 February 1919, 'Papers re war service 1915–1919', Mitchell Library, Sydney, ML MSS 97/4.
24. C.E.W. Bean, letter to his brother, 3 February 1919, AWM38: 3DRL 8042/49.
25. J. Treloar, letter to H.W. Dinning, 6 March 1919, 'AWRS EEF Subsection Cairo reports', AWM25: 1013/1 part 5.
26. C.E.W. Bean, *Gallipoli Mission*, p. 38.
27. G.W. Lambert, journal entry, 14 February 1919, 'Papers re war service 1915–1919', Mitchell Library, Sydney, ML MSS 97/4.
28. Anonymous, *Fusilier Bluff: The experiences of an unprofessional soldier in the Near East, 1918 to 1919*, Geoffrey Bles, London, 1934, pp. 152–53.
29. G.W. Lambert, journal entry, 14 February 1919, 'Papers re war service 1915–1919', Mitchell Library, Sydney, ML MSS 97/4.

CHAPTER 4

1. This 'beetle' had been used in the Suvla Bay landings of August 1915.
2. Captain Cyril Emerson Hughes MBE (1889–1958) enlisted 20 January 1915, 1st Australian Light Horse, 1st Regiment, 3rd Reinforcements, service no. 825; embarked HMAT *Clan MacCorquodale* A6; served on Gallipoli c. 12 May to 25 October 1915, then Middle East; officially transferred to Graves Registration Unit, Gallipoli, on 27 December 1918; discharged from service on 22 October 1919.
3. W.H. James, letter to H.W. Dinning, 11 February 1919, AWM25: 1013/33.
4. W.H. James received his orders from Henry Gullett on 26 November 1918. His tasks were outlined by H.W. Dinning in AWRS Weekly Report, EEF Subsection to London, report no. 5, 6 December 1918, AWM25: 1013/37.
5. The Armistice with Turkey was signed on HMS *Armageddon*, Mudros, 30 October 1918.
6. Report by Lieutenant Colonel J.D. Richardson, 7th Light Horse, cited in H.S. Gullett, *The Australian Imperial Force in Sinai and Palestine*, The Official History of Australia in the War of 1914–1918, vol. VII, Angus & Robertson Ltd, Sydney, 3rd edition, 1936, p. 787.
7. Hughes was attached to the British War Graves Registration Unit and arrived on Gallipoli 10 November 1918.
8. C.E.W. Bean, *Gallipoli Mission*, Australian War Memorial, Canberra, 1948, p. 46.
9. Sapper Arthur Henry Thomas Woolley MSM (c. December 1882 or January 1883 to unknown) first enlisted 14 July 1915 1st Light Horse; discharged 10 January 1915 as medically unfit; re-enlisted 13 April 1917, 1st and 2nd Field Troop Engineers (later 1st Field Squadron), service no. 18040; embarked Sydney on HMAT *Port Sydney* A15; served in the Middle East; transferred to Graves Registration Unit on 27 December 1918; discharged from service on 31 October 1919.
10. J.D. Richardson, *The History of the 7th Light Horse Regiment AIF*, Eric N. Birks, Sydney, c. 1923, p. 109.
11. ibid.
12. C.E.W. Bean, *Gallipoli Mission*, p. 111. Lance Corporal William Henry Spruce (1890–1963) enlisted 28 June 1915, 7th Light Horse, 9th Reinforcements, service no. 1269; embarked Sydney on HMAT *Argyllshire* A8; served in Middle East, transferred to AWRS, Egypt, on 17 January 1919, attached to official photographers; discharged from service on 8 November 1919.
13. G.W. Lambert, journal entry, 18–21 January 1919, 'Papers re war service 1915–1919', Mitchell Library, Sydney, ML MSS 97/4.
14. C.E.W. Bean, *Gallipoli Mission*, p. 130.
15. ibid., p. 127.
16. ibid., pp. 52–53.
17. ibid., p. 46.
18. The Imperial War Graves Commission was formed in May 1917. In 1960 it was renamed the Commonwealth War Graves Commission.
19. Hedley Howe served with the Western Australian 11th Light Horse as a lance corporal on the morning of 25 April 1915.
20. C.E.W. Bean, *Gallipoli Mission*, p. 15. Bean noted that Howe, who had not been well during the expedition, returned to London in February 1919 after thoroughly explaining the events of the landing. *Gallipoli Mission*, p. 113.
21. *Telegraph* [Sydney], 10 May 1915, cited in John Williams, *ANZACS, the Media and the Great War*, Cambridge University Press, Melbourne, 1995, p. 81.
22. ibid.
23. C.E.W. Bean, *Gallipoli Mission*, p. 4.
24. C.E.W. Bean, diary entry, 25 April 1915, AWM38: 3DRL 606/4, p. 8.
25. C.E.W. Bean, *Gallipoli Mission*, p. 79.
26. ibid., p. 207.
27. C.E.W. Bean, handwritten account of the charge at the Nek, in 'Lambert family papers', Mitchell Library, Sydney, ML MSS 97/8, Item 9.
28. G.W. Lambert, journal entry, 17 February 1919, Mitchell Library, Sydney, ML MSS 97/4.
29. ibid.
30. G.W. Lambert, journal entry, 6 March 1919, Mitchell Library, Sydney, ML MSS 97/4.
31. G.H. Wilkins's photograph is AWM G01956; Bean's sketch appears in *Gallipoli Mission*, facing p. 262.

CHAPTER 5

1. C.E.W. Bean, *Gallipoli Mission*, Australian War Memorial, Canberra, 1948, p. 107.
2. C.E.W. Bean, undated typescript sheet describing mapping process used by the Australian Historical Mission, c. 1919, AWM38: 3DRL 8042/48.
3. G.H. Wilkins, letter to C.E.W. Bean, 20 November 1923, AWM38: 6673/468.
4. C.E.W. Bean, 'Australia's records. Preserved as sacred things. Pictures, relics, and writings', *ANZAC Bulletin*, no. 40, 10 October 1917, p. 14.
5. C.E.W. Bean, 'Papers, 1919 related to the Australian Historical Mission', AWM38: 3DRL 8042/56.
6. C.E.W. Bean, *Gallipoli Mission*, p. 99.
7. ibid., p. 59.
8. H. Wellington's bible, since deaccessioned from the Australian War Memorial collection; *Gallipoli Mission*, p. 238.
9. C.E.W. Bean, *Gallipoli Mission*, p. 238.
10. W.H. James frequently called the ANZAC area 'Spooks Plateau'.
11. G.W. Lambert, journal entry, 3 March 1919, 'Papers re war service 1915–1919', Mitchell Library, Sydney, ML MSS 97/4.
12. C.E.W. Bean, 'Preface', *The Story of ANZAC*, The Official History of Australia in the War of 1914–1918, vol. I, 1st edition, 1921, pp. vi–vii.
13. C.E.W. Bean, letter to Defence Secretary T. Trumble, 25 June 1918, AWM16: 4378/1/8.

CHAPTER 6

1. C.E.W. Bean, *Commonwealth of Australia Gazette*, no. 39, 17 May 1915, p. 931.

2 ibid.
3 H.S. Gullett, letter to J. Treloar, 19 May 1918, AWM38: 3DRL 6673/621.
4 Captain Hector William Dinning (1887–1941) enlisted 28 September 1914, Divisional Supply Column Motor Transport, service no. 1997; embarked Melbourne on HMAT *Ceramic* A40; served on Gallipoli c. 25 April to 29 October 1915, then Middle East, France; transferred to AWRS, Cairo, 26 May 1918; discharged from service on 7 July 1920.
5 H.S. Gullett, letter to C.E.W. Bean, 28 November 1917, AWM38: 3DRL 6673/197 part 1.
6 G.W. Lambert, [1918], 'Papers re war service 1915–1919', Mitchell Library, Sydney, ML MSS 97/4, item 6.
7 H.S. Gullett, letter to J. Treloar, 29 July 1918, 'Correspondence between O/C EEF Subsection and AWRS, London', AWM16: 4379/5/21.
8 Honorary Lieutenant James Pinkerton Campbell (1869–1935) enlisted 12 September 1914, 8th Light Horse Regiment (later Australian Army Pay Corps), service number 193; embarked Melbourne on HMAT *Star of Victoria* A16; served on Gallipoli, 16 May to 13 August 1915; injured and transferred to Australian Army Pay Corps, then AWRS, Cairo, as official photographer from 13 March 1918; discharged from service on 18 April 1919.
Lieutenant Oswald 'Ossie' Hillam Coulson MSM (1872–1945) enlisted 1 April 1916, 1st Australian Flying Squadron, 2nd Reinforcements, service no. 455; embarked Melbourne on RMS *Malwa*; served in Middle East with 67th Squadron and 1st Squadron, AFC, then AWRS, Cairo, as official photographer from 9 November 1918; discharged from service on 2 March 1920.
9 The Royal Air Force was formed on 1 April 1918 by amalgamating the Royal Flying Corps and the Royal Naval Air Service.
10 Cablegram, 2 January 1919, in response to C.E.W. Bean's request of 26 November 1918, AWM38: 3DRL 8042/48.
11 Cablegram, ADMINAUST London to STRALIS Cairo, 25 November 1918, AWM16: 4359/1/13.
12 W.H. James, diary entry, 26 November 1918, typescript of W.H. James's diary.
13 H.W. Dinning, 'The future War Museum' in *Kia Ora Coo-ee News*, 11 September 1918, no. 4, p. 1.
14 H.W. Dinning, AWRS Weekly Report, EEF Subsection, Cairo, report no. 5, 6 December 1918, AWM25: 1013/37.
15 ibid. Several of Dinning's statements are misleading: the big guns at Gaba Tepe did not in fact produce enfilading fire and, although concealed in the general area of a place called 'The Olive Grove', they were actually surrounded by low oak trees.
16 Captain William Siborne's detailed battlefield model of the 1815 battle of Waterloo was first displayed at the United Services Institute in 1838 and is now on view in the National Army Museum, London. The Military Museum, Istanbul, also has on display a model of the Gallipoli battlefields showing Turkish resistance to the British and allied invasion of 1915.
17 H.W. Dinning, AWRS Weekly Report, EEF Subsection, Cairo to London, report no. 5, 6 December 1918, AWM25: 1013/37.
18 ibid.
19 Biographical information regarding William Swanston has been drawn from his service record, RAF no. 17582, National Archives, London, AIR79/196, and information contributed by a family descendant to Peter Spratt's Edinburgh photography website, <http://www.edinphoto.org.uk>.
20 H.W. Dinning, AWRS Weekly Report, EEF Subsection, Cairo to London, report no. 5, 6 December 1918, AWM25: 1013/37.
21 ibid.
22 See C.E.W. Bean, *The Story of ANZAC*, The Official History of Australia in the War of 1914–1918, vol. II, Angus & Robertson, Sydney, 1924, pp. 606–31 for a detailed account of this day.
23 W.H. James, letter to H.S. Gullett, 5 January 1919, AWM40: 48.
24 W.H. James, letter to H.W. Dinning, 5 January 1919, AWM25: 1013/33.
25 James's friend was C.W. Gibson, service no. 195, from A Squadron, 1st Australian Light Horse Regiment. W.H. James, letter to H.W. Dinning, 5 January 1919, AWM25: 1013/33.
26 ibid.
27 W.H. James, letter to H.S. Gullett, 17 January 1919, AWM40: 48.
28 ibid.
29 ibid.
30 W.H. James, letter to H.W. Dinning, 17 January 1919, AWM25: 1013/33.
31 After surveying the area with Major Zeki, Bean came to a similar conclusion about the site of the 'Beachy Bill' guns.
32 C.E.W. Bean, *Gallipoli Mission*, Australian War Memorial, Canberra, 1948, p. 280.
33 G.W. Lambert, journal entries, 24 February 1919 and 4 March 1919, 'Papers re war service 1915–1919', Mitchell Library, Sydney, ML MSS 97/4.
34 G.W. Lambert, journal entry, 7 March 1919, 'Papers re war service 1915–1919', Mitchell Library, Sydney, ML MSS 97/4.
35 G.W. Lambert, journal entry, 6 March 1919, 'Papers re war service 1915–1919', Mitchell Library, Sydney, ML MSS 97/4.
36 C.E.W. Bean, *Gallipoli Mission*, p. 283.
37 C.E.W. Bean, diary entry, 7 March 1919, AWM38: 3DRL 606/231/1, pp. 36–37.
38 C.E.W. Bean, diary entry, 6 May 1915, AWM38: 3DRL 606/7A, p. 5.
39 C.E.W. Bean, *Gallipoli Mission*, p. 286.
40 ibid.
41 C.E.W. Bean, *The Story of ANZAC*, vol. II, p. 40. Bean revealed that he was this soldier in *Gallipoli Mission*, p. 298.
42 These recommendations can be found on the Australian War Memorial website, <http://www.awm.gov.au/honours/awm28/index.asp>.

CHAPTER 7

1 Hedley Howe, who had not been well during the expedition, returned to London after a short stay on Gallipoli.
2 G.W. Lambert, journal entry, 10 March 1919, 'Papers re war service 1915–1919', Mitchell Library, Sydney, ML MSS 97/4.
3 C.E. Hughes, letter to C.E.W. Bean, 25 April 1920, 'Relics from Gallipoli Battlefield. Lt Col Hughes (1919–1937)', AWM93: 7/4/167a.
4 Much of the material collected in 1919 remains in the Memorial's collection, including one of the lifeboats that was eventually salvaged from the beach, RELAWM0586.001.

5. C.E.W. Bean, *Gallipoli Mission*, Australian War Memorial, Canberra, 1948, p. 310.
6. ibid., p. 312.
7. G. Hunter Rogers, *My First Eighty Years*, p. 73.
8. G.W. Lambert, journal entry, 14 March 1919, 'Papers re war service 1915–1919', Mitchell Library, Sydney, ML MSS 97/4.
9. C.E.W. Bean, *Gallipoli Mission*, p. 312.
10. ibid., p. 313.
11. ibid., p. 313.
12. G.W. Lambert, journal entry, 15–18 March 1919, 'Papers re war service 1915–1919', Mitchell Library, Sydney, ML MSS 97/4.
13. C.E.W. Bean, *Gallipoli Mission*, p. 313.
14. ibid., p. 317.
15. ibid., p. 317
16. ibid., p. 318.
17. Lance Corporal Gerald Calcutt (1890–1915) enlisted 15 August 1914, 7th Battalion, service no. 474; embarked Melbourne on HMAT *Hororata* A20; served on Gallipoli, 25 April to 24 May 1915; posted as missing in action; Court of Enquiry, 11 November 1916, confirmed he was killed in action on 24 May 1915.
18. C.E.W. Bean, *The Story of ANZAC*, vol. II, pp. 662–63.
19. Private Brendan Calcutt (1895–1916) enlisted 7 April 1915, 14th Battalion, 6th Reinforcements, service no. 2124; embarked Melbourne on HMAT *Wandilla* A62; served on Gallipoli, 1–8 August 1915; reported wounded and taken prisoner of war on 8 August 1915; died on 18 December 1916 (death confirmed by Ottoman Red Crescent Society, 5 September 1917).
20. General Henry 'Harry' George Chauvel GCMG KCB (1865–1945) commissioned, 1886, in Upper Clarence Light Horse; resigned and, in 1890, re-commissioned in Queensland Mounted Infantry (QMI), then as Captain, Queensland Permanent Military Forces; served with QMI in South Africa, 1899–1900; continued to serve with AMF; during First World War, commanded 1st Light Horse Brigade at Gallipoli and the Desert Mounted Corps in the Middle East; from 1919, Inspector General for Army to retirement 1930, also Chief of Staff and appointed General; during the Second World War, Inspector General of the Volunteer Defence Corps until his death on 4 March 1945.
21. Lambert made a pencil portrait of Chauvel on 15 February 1918, AWM ART02734.
22. C.E.W. Bean, *Gallipoli Mission*, p. 321.
23. Emir Feisal (1883–1933) was born in Saudi Arabia and sided with Great Britain during the war. He worked with T.E. Lawrence (Lawrence of Arabia) to overthrow the Ottoman Empire. In 1918 he became part of a new Arab government and in 1919 he led the Arab delegation to the Paris Peace Conference. For a short period he was King of Greater Syria and then from 1921 to 1933 he was King of Iraq.
24. C.E.W. Bean, *Gallipoli Mission*, p. 321
25. G. Hunter Rogers, *My First Eighty Years*, p. 74.
26. ibid.
27. C.E.W. Bean, *Gallipoli Mission*, p. 323.
28. W.H. James noted in his diary that Swanston did some printing for Wilkins, 6 March 1919.
29. This set of photographs is in the AWM collection, AWM38: 3DRL 6673/1016.
30. Service records for Buchanan and Rogers say that they departed on the *Kildonan Castle*, 1 April 1919, service records for Balfour and Bazley cite 2 April 1919 as the departure date, as does C.E.W. Bean in *Gallipoli Mission*, p. 324.
31. H.W. Dinning, AWRS Weekly Report, EEF Subsection, Cairo, report no. 13, 15 April 1919, AWM25: 1013/37.
32. W.H. Swanston service record, National Archives, London, AIR79/196.
33. G.W. Lambert, journal entry, 24 June 1919, 'Papers re war service 1915–1919', Mitchell Library, Sydney, ML MSS 97/4.
34. G.W. Lambert, undated note but c. 26 July 1919, 'Papers re war service 1915–1919', Mitchell Library, Sydney, ML MSS 97/4.
35. G.W. Lambert, journal entry, 7 June 1919, 'Papers re war service 1915–1919', Mitchell Library, Sydney, ML MSS 97/4.
36. G.W. Lambert, journal entry, 9 July 1919, 'Papers re war service 1915–1919', Mitchell Library, Sydney, ML MSS 97/4.

CHAPTER 8

1. H.W. Dinning, AWRS Weekly Report, EEF Subsection, Cairo to London, report no. 13, 15 April 1919, AWM25: 1013/37.
2. When it opened its displays in Melbourne in 1922, the institution was called the Australian War Museum. The name was officially changed to the Australian War Memorial in 1925. To prevent confusion, the term Australian War Memorial is used throughout this book.
3. The estimate of about 150 images is based on a record W.H. James kept of plates exposed each day during the time he was on Gallipoli.
4. C.E.W. Bean, letter to parents, 6 July 1919, AWM 38: 3DRL 7447/7, p. 4.
5. A.W. Bazley, 'Australia's Official History of World War I', *Stand To*, vol. 6, no. 6, November 1958 to January 1959, p. 29.
6. C.E.W. Bean, Preface to first edition, reprinted in *The Story of ANZAC*, p. xxix.
7. ibid.
8. For an interesting account of Bean's methods regarding 'evidence', see C.E.W. Bean, 'A war historian's experience with eyewitnesses', *Jadunath Sarkar Memorial Volumes*, Panjab University, India, 1958.
9. Out of the total of 55 photographic illustrations in the first volume of the official history, Bean incorporated 32 of his 1915 images with 17 photographs taken by Hubert Wilkins in 1919. The second volume, printed in 1924, benefited from the growing number of donated photographs but still had 33 images taken by Bean in 1915 out of a total of 99 illustrations.
10. C.E.W. Bean, Preface to first edition, reprinted in *The Story of ANZAC*, The Official History of Australia in the War 1914–1918, vol. I, Sydney, 3rd edition, 1934, p. xxix.
11. This album is AWM38: 3DRL 606/1016.
12. Lieutenant Colonel Noel Medway Loutit DSO and Bar (1884–1983) enlisted 18 September 1914, 10th Infantry Battalion (later rose to command 45th Battalion); commissioned as 2nd Lieutenant; embarked Adelaide on HMAT *Ascanius* A11; served on Gallipoli 25 April to c. 19 December 1915, then Middle East and Western Front; discharged from service on 9 January 1920.

13 C.E.W. Bean, *The Story of ANZAC*, vol. I, reproduced facing p. 347.
14 ibid., p. xii.
15 RELAWM09636.
16 See Michael McKernan, *Here Is Their Spirit: A history of the Australian War Memorial 1917–1990*, University of Queensland Press and the Australian War Memorial, St Lucia, 1991, p. 68, and *Descriptive Catalogue of Exhibition of Enlargements: Official war photographs*, Australian War Museum, Melbourne, 1921.
17 *Descriptive Catalogue of Exhibition of Enlargements, Official War Photographs*, cover text.
18 Charles Bean, G01294, Hubert Wilkins, G02018.
19 C.E.W. Bean, *Photographic Record of the War*, The Official History of Australia in the War of 1914–1918, vol. XII, first published in Sydney, 1923; contains 753 photographic illustrations.
20 'War trophies exhibition. Preparations for Melbourne display', *The Age*, Melbourne, 2 April 1921, p. 18.
21 A full account of the dioramas can be found in Laura Back and Laura Webster, *Moments in Time: Dioramas at the Australian War Memorial*, New Holland, Sydney, 2009.
22 G.W. Lambert, 'Autobiography of George Lambert', Mitchell Library, Sydney, ML MSS A1811, p. 75.
23 Cited in Amy Lambert, *The Career of G.W. Lambert, A.R.A: Thirty years of an artist's life*, Society of Artists, Sydney, 1938, reprinted by Australian Artist Editions, Sydney, 1977, p. 135.
24 Alexander Colquhoun, 'Battles in oils. Artists depict the spirit of Anzac. Fine show at War Museum', *The Herald*, Melbourne, 4 May 1922.
25 Advertisement in *Descriptive Catalogue: Official war photographs*, Australian War Museum, Melbourne, c. 1922, p. 30.
26 *The Sun News-Pictorial*, Melbourne, 25 April 1924, p. 1.
27 G.W. Lambert, journal entry, 16 February 1919, 'Papers re war service 1915–1919', Mitchell Library, Sydney, ML MSS 97/4. Lambert's statement that all 'went down to a man' is a little misleading: although a horrific action, there were survivors. Of the 600 men who took part, 234 were killed and some 140 wounded. See Peter Burness, *The Nek: The tragic charge of the Light Horse at Gallipoli*, Kangaroo Press, Kenthurst, NSW, 1996, for the definitive account of this action.
28 Lambert's drawing *First geometric design for 'The Nek'* is on the back of *Jerusalem from the top of the Dung Gate*, ART02855. The sketch probably dates from late March 1919.
29 'Lambert's life goes on. Famous canvases on show to-day', *Daily Guardian*, Sydney, 26 November 1930, p. 8.
30 G.W. Lambert, letter to C.E.W. Bean, 5 March 1924, AWM38: 3DRL 6673 Part 302.
31 G.W. Lambert, letter to C.E.W. Bean, early 1924, AWM38: 3DRL 6673 Part 303.
32 Senator G. Pearce, Chairman of the Australian War Museum Art Committee, draft letter to G.W. Lambert, AWM38: 3DRL 6673 Part 302.
33 *Sydney Morning Herald*, 11 September 1924, p. 8 and *Table Talk*, 18 September 1924, p. 14.
34 Amy Lambert, *The Career of G.W. Lambert, A.R.A: Thirty years of an artist's life*, p. 153.
35 Stanley Kingsbury, 'Temple of memories. Message of the War Museum', *Evening News*, Sydney, 7 April 1925, p. 10.
36 'Relics of the war: Museum is a shrine to thousands', *Daily Telegraph*, Sydney, 11 July 1933.
37 ibid.
38 *Guide to Australian War Memorial*, Australian War Memorial, Canberra, 1941.
39 This is the steel lifeboat from HMT *Ascot* A33 used in the landings on Gallipoli by No 13 Battalion, AIF, RELAWM0586.001.

EPILOGUE

1 'John Balfour: Important army job', *Reveille*, 1 March 1945, p. 3.
2 C.E.W. Bean, *Gallipoli Mission*, p. vii.
3 G.W. Lambert, letter to C.E.W. Bean, 19 January 1924, AWM38: 3DRL 6673, Item 467.
4 C.E.W. Bean, 'Sir Hubert Wilkins', *Reveille*, 1 September 1933, pp. 7, 33.

SELECT BIBLIOGRAPHY

Descriptive Catalogue of Exhibition of Enlargements: Official war photographs, Australian War Museum, Melbourne, c. 1921.

Australian War Museum: The relics and records of Australia's effort in defence of the Empire 1914–1918, Australian War Museum, Melbourne, 1922.

C.E.W. Bean, 'Australia's records. Preserved as sacred things. Pictures, relics, and writings', *Anzac Bulletin*, vol. 40, 10 October 1917 [written 29 Sept 1917].

C.E.W. Bean, *The Story of ANZAC*, The Official History of Australia in the War of 1914–1918, vol. I, Angus & Robertson, Sydney, 1921 [3rd edition, 1934].

C.E.W. Bean, *The Story of ANZAC*, The Official History of Australia in the War of 1914–1918, vol. II, Angus & Robertson, Sydney, 1924.

C.E.W. Bean, *Photographic Record of the War*, The Official History of Australia in the War of 1914–1918, vol. XII, Angus & Robertson, Sydney, 1923.

C.E.W. Bean, *Gallipoli Mission*, Australian War Memorial, Canberra, 1948, reprinted 1952 and published again in association with ABC Enterprises in 1990.

Harvey Broadbent, *Gallipoli: The fatal shore*, Penguin Books, Camberwell, Victoria, 2005.

Peter Burness, *The Nek: The tragic charge of the Light Horse at Gallipoli*, Kangaroo Press, Kenthurst, NSW, 1996.

Les Carlyon, *Gallipoli*, Macmillan, Sydney, 2001.

Kevin Fewster, *Bean's Gallipoli: The diaries of Australia's official war correspondent*, Allen & Unwin, Sydney, 2007.

Paul Fussell, *The Great War and Modern Memory*, Oxford University Press, Oxford and New York, 1975.

Bill Gammage, *The Broken Years: Australian soldiers in the Great War*, Penguin Books, Ringwood, 1975.

Janda Gooding, '"Beauty in Hell": George Lambert in Gallipoli and Palestine', *Wartime*, 38, April 2007, pp. 62–65.

Anne Gray, *Art and Artifice: George Lambert 1873–1930*, Craftsman House, Sydney, 1996.

Anne Gray, *George W. Lambert Retrospective: Heroes and icons*, National Gallery of Australia, Canberra, 2007.

John Grierson, *Sir Hubert Wilkins: Enigma of exploration*, Robert Hale Ltd, London, 1960.

Samuel Hynes, *A War Imagined: The First World War and English culture*, The Bodley Head, London, 1990.

Robert Rhodes James, *Gallipoli*, B.T. Batsford Ltd, London, 1965.

Shaune Lakin, *Contact: Photographs from the Australian War Memorial collection*, Australian War Memorial, Canberra, 2006.

Amy Lambert, *The Career of G.W. Lambert, A.R.A: Thirty years of an artist's life*, Society of Artists, Sydney, 1938, reprinted by Australian Artist Editions, Sydney, 1977.

Dudley McCarthy, *Gallipoli to the Somme: The story of C.E.W. Bean*, John Ferguson, Sydney, 1983.

Michael McKernan, *Here is Their Spirit: A history of the Australian War Memorial, 1917–1990*, University of Queensland Press and the Australian War Memorial, St Lucia, 1991.

Sue Malvern, *Modern Art, Britain and the Great War*, Paul Mellon Centre for Studies in British Art, Yale University Press, New Haven and London, 2004.

John Masefield, *Gallipoli*, William Heinemann, London, 1916.

Ann Millar, 'Gallipoli to Melbourne: The Australian War Memorial, 1915–19', *Journal of the Australian War Memorial*, no. 10, April 1987, pp. 33–42.

Alan Moorehead, *Gallipoli*, Hamish Hamilton, London, 1956.

George L. Mosse, *Fallen Soldiers: Reshaping the memory of the world wars*, Oxford University Press, Oxford and New York, 1990.

Simon Nasht, *The Last Explorer: Hubert Wilkins: Australia's unknown hero*, Hodder Australia, Sydney, 2006.

J.D. Richardson, *The History of the 7th Light Horse Regiment AIF*, Eric N. Birks, Sydney, c. 1923.

G. Hunter Rogers, *My First Eighty Years*, self-published, Melbourne, c. 1982.

Bruce Scates, *Return to Gallipoli: Walking the battlefields of the Great War*, Cambridge University Press, Melbourne, 2006.

Nigel Steel and Peter Hart, *Defeat at Gallipoli*, Macmillan, London, 1994.

Lowell Thomas, *Sir Hubert Wilkins: His world of adventure: An autobiography*, Readers Book Club in association with The Companion Book Club, Melbourne, 1963.

Richard White, *Inventing Australia: Images and identity 1688–1980*, Allen & Unwin, NSW, 1981.

George Hubert Wilkins, *Australian War Photographs: A pictorial record from November 1917 to the end of the war*, AIF Publications section, London, 1919.

Lola Wilkins (ed.), *Artists in Action: From the collection of the Australian War Memorial*, Australian War Memorial, Canberra, 2003.

John F. Williams, *Quarantined Culture: Australian reactions to modernism 1913–1939*, Cambridge University Press, Melbourne, 1995.

Denis Winter, *Making the Legend: The war writings of C.E.W. Bean*, University of Queensland Press, St Lucia, 1992.

Bart Ziino, *A Distant Grief: Australians, war graves and the Great War*, University of Western Australia Press, Crawley, 2007.

GLOSSARY OF ACRONYMS

AFC
Australian Flying Corps

AIF
Australian Imperial Force, the all-volunteer force raised in 1914 for overseas service in the First World War

ANZAC
Australian and New Zealand Army Corps. The formation into which troops of both countries were grouped for the Gallipoli landings of 1915

AWM
Australian War Memorial. This name was established by the *Australian War Memorial Act*, 1925. (While at the Exhibition Building in Melbourne, from 1922 to 1925, named the Australian War Museum; and while at the Exhibition Building in Sydney, from 1925 to 1935, named the Australian War Memorial Museum.)

AWRS
Australian War Records Section. Formed in May 1917 to organise the collection of war records and relics and the official art and photography schemes

HMAS
His/Her Majesty's Australian ship

HMS
His/Her Majesty's ship

RAF
Royal Air Force, prior to April 1918 known as the Royal Flying Corps (RFC)

RAN
Royal Australian Navy

CHRONOLOGY OF THE AUSTRALIAN HISTORICAL MISSION

1918

30 October — Armistice signed by Ottoman Empire (Turkey from 1922), Britain and its allies.

10 November — Lieutenant Cyril Hughes of the 1st Field Squadron, Australian Engineers, landed with the British Graves Registration Unit on Gallipoli to survey and consolidate cemeteries and locate the graves of Australian soldiers.

11 November — Hostilities on the Western Front ceased.

12 November — Charles Bean travelled from France to London to seek approval to revisit Gallipoli.

25 November — Henry Gullett, head of the Australian War Records Section (AWRS), Cairo, was advised that Charles Bean, accompanied by a photographer and artist, would visit Gallipoli in early 1919. Gullett immediately appointed Lieutenant William James to proceed to Gallipoli to collect relics and make a photographic survey.

3 December — The Australian Government was notified by Hughes that there had been some desecration of Australian graves on Gallipoli. It was assumed this damage had been done by Turkish soldiers.

4 December — A detachment of Australians from the 7th Light Horse arrived at Gallipoli, and stayed until 10 January 1919. Several soldiers were seconded to work with the AWRS photographers and the Historical Mission.

19 December — Official approval from the Australian Government was received for Bean and his party to travel to Gallipoli.

30 December — Lieutenant William James and Sergeant William Swanston from the AWRS, Cairo, arrived on Gallipoli.

1919

11 January — Final confirmation of the personnel who would form the Australian Historical Mission was received: Charles Bean (historian), Lieutenant John Balfour (records and Bean's assistant), Staff Sergeant Arthur Bazley (records), Lieutenant Herbert Buchanan (mapping and topographical expert), Staff Sergeant George Rogers (mapping assistant), Captain George Wilkins (photographer), George Lambert (artist, promoted to captain on 17 January 1919) and Lieutenant Hedley Howe (a veteran of the 1915 landing).

18 January — The Historical Mission left Waterloo Station, London, by train for Southampton.

19 January — The Historical Mission sailed on *City of Poona* to Le Havre then travelled by train to Paris, arriving the next day. They met with Gullett and other Australians who were attending the Paris Peace Conference before leaving Paris for Genoa.

22 January– 7 February — James and Swanston took leave to visit Constantinople.

22–24 January — The Historical Mission rested for a few days in Rome. Bazley became seriously ill with influenza and was forced to withdraw from the Historical Mission, later joining them in Cairo.

25–26 January — Stayed at Taranto rest camp.

26 January — Embarked on an old British naval sloop, *Asphodel*, for an overnight voyage to Malta. They spent three days (27–29) in Valletta, at the Hotel Santa Lucia, and also went to the opera *Fedora*.

30 January — Left Malta for Suda Bay, Crete, on *Princess Ena*, arriving on 1 February.

2 February — Left Crete for Lemnos on *Princess Ena*. During the voyage, Bean gave a lecture to the group on the importance of Gallipoli.

3 February — Arrived at Mudros Harbour, Lemnos.

5 February — Embarked on *Princess Ena* and arrived the next day at Chanak.

6–14 February — Lambert, Balfour, Howe, Buchanan and Rogers stayed at Chanak while Bean and Wilkins went on to Constantinople to interview officers and make further arrangements about the work of the Historical Mission. While in Constantinople, Bean and Wilkins met up with William James from the AWRS, Cairo.

CHRONOLOGY OF THE AUSTRALIAN HISTORICAL MISSION

14 February	Bean and Wilkins returned to Chanak and, with the rest of the group, crossed over to Kilid Bahr, a fort on the Gallipoli peninsula. They stayed the first night at Cham Burnu, an old Turkish hospital camp.
15 February	The group travelled across the peninsula and made camp at Legge Valley (behind Lone Pine and the old Turkish lines). Bean, Wilkins and Lambert rode along the ridges to Lone Pine, Quinn's Post, Shrapnel Gully and then down to ANZAC Cove.
16 February	Bean inspected the cemeteries and gravesites with Hughes from the Graves Registration Unit. Overnight he prepared a preliminary report on the state of the graves that was cabled to London reassuring authorities that damage to gravesites was not widespread. James assisted Wilkins to construct a makeshift darkroom out of a salvaged water tank.
17 February– 6 March	The Historical Mission criss-crossed the ANZAC area, making surveys, sketch maps, collecting relics and documenting the landscape through photographs and paintings.
17 February	The group followed the route taken by Hedley Howe on the morning of 25 April 1915.
20 February	Bean crossed to Chanak to await the arrival of a Turkish officer, Zeki Bey, who had been assigned to assist in the work of the Historical Mission. Hedley Howe, who had been unwell, probably left for London at this time.
21 February	Zeki Bey arrived and stayed with the Historical Mission for seven days. The following day the whole group went to Hill 60.
27 February	The group visited Gaba Tepe to investigate the position of gun emplacements and what the Turks could have seen of the Australian positions.
28 February– 3 March	The weather deteriorated, and a blizzard interrupted all work on 3 March.
7 March	Bean, Lambert, Wilkins and Buchanan rode down to the Helles area. Balfour and Rogers remained at the camp to pack up and follow later in the day.
8–9 March	The group spent two days working around Krithia and Tommies' Trench and traced troop movements of 8 May 1915.
10 March	The Historical Mission left Gallipoli and travelled to Constantinople on the Greek steamer *Spetsai*.
11–13 March	In Constantinople the group stayed at the Officers' Rest House. They made travel arrangements and prepared for their transportation in enclosed cattle rail wagons to Cairo.
14 March	Crossed to Scutari, then travelled by rail through Turkey.
15–18 March	Passed through Konia, the Taurus Mountains, the Cilician Gates and then down onto the plain and on to Adana.
19–22 March	Travelled from Adana to Bozanti in the Amanus Ranges before heading south to Aleppo, where they met General Henry Chauvel.
23 March	Having travelled through the valleys past Baalbek to Ryak Junction, the Historical Mission arrived in Damascus, where they met T.E. Lawrence (Lawrence of Arabia). They stayed overnight in a hotel as a break from sleeping in the rail wagons before continuing their journey.
24 March	Arrived at Jerusalem late in the day.
25 March	The group spent the day in Jerusalem before continuing their journey overnight to Cairo.
26 March– 2 April	The Historical Mission was in Cairo, where Arthur Bazley rejoined them.
2 April	Bean, Balfour, Bazley, Rogers and Buchanan left Cairo for Kantara to await their departure for Australia on 8 April on the *Kildonan Castle*. Lambert was hospitalised in Cairo with dysentery, malaria and heart irregularities.
12 April	Wilkins left to continue his photographic work on the Western Front battlefields. James and Swanston returned from Gallipoli to Cairo with the relics that had been collected.
31 May	Lambert was discharged from hospital and continued to work in the Middle East, making sketches and paintings of the activities of the Australian Light Horse until August.
2 August	George Lambert left Port Said for Britain on board HT *Caledonia*.

INDEX

Achi Baba *102*, 125, 154, 159, 162
Achi Baba, from Tommy's [sic] Trench, Helles (George Lambert) 125, *160*
Aegean Sea 50, *90*
Aleppo 178
Aleppo citadel (George Lambert) *180*
Anafarta 133
ANZAC, from Gaba Tepe (George Lambert) *86*, 87
ANZAC, the landing, 1915, (George Lambert) 80, 202, *204–5*, 209, 214
ANZAC Cove 11, 70, *73*, 148, *150*
ANZAC Cove, Gallipoli (George Lambert) *72*
ANZAC landing 74–9
Ari Burnu 74, *75*, 116, *116*, 152, 209
Asmak Dere *147*
Australian Flying Corps 25
Australian Historical Mission vi, 1, 21, 26, 29, 43, 50, 55, 57, 65, 79, 108, 136
  achievement 215–7
  arriving at Gallipoli 65–6
  arriving in Cairo 182
  battlefield evidence 189–90
  to Cairo via the Taurus Mountains 165–78
  camping in Legge Valley 69
  collaboration with the AWRS 148
  departing for Australia 182
  disembarking at Kilid Bahr 67
  following the steps of the landing 74
  on Hill 60 87, 218
  leaving for Cairo *164*
  leaving Gallipoli 165
  moving to Cape Helles 154
  at the Nek 105
  objectives xi, 2–3, 38–41, 93
  paintings and photographs 189–90
  photographs available to the public 198
  role of photography 117
  route through Europe to the Middle East 1919 44, *44*
  work routine on Gallipoli 70

Australian Imperial Force (AIF) x–xi
  at Montauban, France 1916 *24*
Australian Light Horse 29
  at Beersheba 38, 43
Australian Light Horseman (alternative title *Digger*), (George Lambert) *33*
Australian POWs on the Taurus Railway 177, *177*
Australian War Memorial ix, xi, 4
  Melbourne exhibition 198
  Orientation Gallery 215
Australian War Museum
  Melbourne Exhibition Buildings *188*
Australian War Records Section (AWRS) xi, 14, 17, 25, 26, 38, 45, 47, 93, 95, 165, 182
  Cairo photography project 127–8, 133, 182
  cataloguing the landscape 148
  collecting battlefield material *190*
  darkroom, Cairo *129*
  photographic project 131
  war diary room *183*
Azak Dere 142

Baalbek ruins *173*
Baby 700 *95*, *107*, 108, *111*, 148, *149*
Balcony of trooper's ward, 14th Australian General Hospital, Abbassia (George Lambert) *186*
Baldwin, Herbert viii
Balfour, Lieutenant John 38, 45, 47, 70, 79, 93, 164
  biographical note 225n23
  later life 220
  official war histories 193
battlefields, relief models 133
Battleship Hill 105, 108, *111*, *135*, 162
Bazley, Staff Sergeant Arthur 38, 45, 47, 182, *191*
  biographical note 225n22
  later life 219–20
  official war histories 193
Beachy Bill (Turkish artillery) 90, 133, 142
  likely location *147*

Bean, Effie (wife of Charles) 194
Bean, Honorary Captain Charles viii, ix, 5, 10, 14, 17, 21, 22, *24*, 25, 26, 38, 41, 50, 54, 57, 65, 66, 70, 74, 79, 91, 92–3, 100–5, 112, 125, 127, 128, 133, 148, 154, 162, 165, 178, *192*, *194*, 215
  annotates Wilkin's photographs *199*, *200*, *201*
  biographical note 223n4 (Intro)
  collection methodology 93
  embarks at Port Melbourne *5*
  on Gallipoli, 1915 *6*
  at Gueudecourt, France, 1917 *30*
  at his Victoria Barracks office, Sydney *196*
  later life 218
  at Martinpuich, France, 1917 *40*
  official war correspondent x
  official war histories 191
  record of Taurus rail journey 169–77
  at sea in 1914 *8*
  style of reporting 7–10
  at Tuggeranong Homestead 193
  uses photographs to map the first day's advance 197–8
Bean, the Reverend Edwin (Charles's father) *5*
Beersheba, Palestine 128
Behind the Turkish lines, Gallipoli (George Lambert) *64*
Benson, Lieutenant George 1
  biographical note 224n21
Big gun emplacement, Fort of Chanak (George Lambert) 59, *61*
Blackburn, Private Arthur 198
Bloody Angle 139
Boer War 4
Boxer Rebellion (China) 4
Bozanti 169
Bridges Road 97
British Empire and Australia 4–5
British Graves Registration Unit 41
Brown, Private (draughtsman) *191*
Brown's Dip 94
  cemeteries 155
Buchanan, Lieutenant Herbert 39, 45, 70, 79, 93, 100, 102, 104–105, *106*, *120*, 154, 162, 164, 220
  biographical note 225n25
Burnt Gully, Gallipoli (George Lambert) *89*

Calcutt, Private Brendan 177
  biographical note 228n19
Calcutt, Lance Corporal Gerald 177
  biographical note 228n17
Campbell, Honorary Lieutenant James 128
  biographical note 227n8
Canadian War Records Office 14
Cape Helles 50, 57, 154, 159, 160, *161*, 162
Captain Wilkins on the Hindenburg Line, (Will Dyson) *29*
Carter, Miss Eleanor Maida 46
Casualty Corner *141*, 142, 165
Chailak Dere *126*
Cham Burnu 136
Chanak (now Çannakale) 50, *52*, 55, 56–7, 59, 65, *135*, *145*, 162
Charles E.W. Bean (George Lambert) *216*
Chatham's Post *140*
Chauvel, General Harry *34*, 178
  biographical note 228n20
Chunuk Bair 108, 112, 162
Çimenlik 59
Clarke's Gully 142
Constantinople (now Istanbul) *42*, 45, *54*, 55, *57*
Cooee Gully *121*
Cook, Right Honourable Sir Joseph 46
Corrigan, Mr W.E. 46
Coulson, Lieutenant Oswald 128, 182
  biographical note 227n8
Courtney's Post 139
cover illustration for *Australian War Photographs* 203, *203*

Damascus 178
Dardanelles 50, *52*, *52*, 55, 59, *135*, *145*, 154
  British occupation of forts 65

INDEX

Dardanelles from Chanak, effects of blizzard on Gallipoli (George Lambert) 59, *62*
De Knoet Farm (on Broodseinde Ridge) 1917 *31*
Dead Man's Ridge 139, *144*
Deane, Lieutenant P.E. 46
Dinning, Captain Hector 128, 133, 182
  biographical note 227n4
dioramas 203
Dyson, Honorary Lieutenant William (artist) 17, 29
  biographical note 224n31

Eggleston, Lieutenant F.W. 46
Egypt 38, 128
8th (Victorian) Regiment
  at the Nek 79
El Arish, (George Lambert) *36*
El Arish, Egypt 38
11th Battalion 38
Elsa Trench *28*

Feisal, Emir 178
  biographical note 228n23
Figure study for 'The charge of the 3rd Light Horse Brigade at the Nek, 7 August 1915' (George Lambert) *214*
First Balkan War 25, *26*
1st ANZAC Headquarters in France 100
1st Australian Division Headquarters 38
1st Division staff at the landing *13*
1st Light Horse Brigade 139, 142
4th Australian Brigade *8*
  at Hill 60 112
Fullwood, Lieutenant A. Henry 43
  biographical note 225n2

Gaba Tepe 70, 82, 87, *88*, 90, *90*, 98, 112, 142, *147*
Gallipoli 9, 10
  1919 (map) *132*
  casualty rate 10
  landscape 148
  recording the landscape 127
  wildflowers *84*
Gallipoli (illustration from *The ANZAC Book*) *2*
Gallipoli, from the Chanak side (George Lambert) 59, *63*

Gallipoli Mission (a book by Charles Bean) 1, 2, 219
Gallipoli souvenirs
  postcards *10*, *19*
  printed scarf *3*
Gallipoli wild flowers (George Lambert) *82*
Gammage, Bill (historian) 21
The Gap *153*
Garran, Sir Robert 46
German Officers' Trench 70, *101*, 105, *106*
Graves Registration Unit 65, 66, 74, 93, 94, 112, 136, 155, 162
Gullett, Captain H.S. 46, 47, 128, 133
  biographical note 225n17
Gun Ridge 198

Hadschkiri 169, 177
Head of a Turk, Chanak (George Lambert) *57*
Hell Spit *74*, 162
Helles Beach *124*
Hill 60 70, 112, *114*
Hill 100 *114*
Hill 971 *123*, *145*, 153
HMAS *Australia* *12*
HMAS *Parramatta* *54*
HMS *Grafton* 10
HMT *Kildonan Castle* *191*
Howe, Lieutenant Hedley 38, 41, 45, 74, 75, 79, 80, 202
  biographical note 225n24
  later life 220–1
Hughes, Captain Cyril 65, 66, 74, *94*, 136, 165
  biographical note 226n2
Hughes, Prime Minister W.M. ('Billy') 11, *12*, 46, 47, 215
Hurley, Honorary Captain Frank viii, 14, *20*, *23*
  biographical note 224n28
  composite photographs 22–3
  recording the Light Horse units 128

Imbros *91*
Imperial War Graves Commission 74, 165
Inside the Fort, Chanak (George Lambert) *58*
Islahiye, Taurus Mountain railways *170*, *171*

James, Lieutenant William xi, 4, 65, 70, 93, 100, 104, 127, 128, 131, 133, 138–9, 142, 144, 148, 154, 155, 165, 215
  biographical note 223n8
  later life 221
  partnership with William Swanston 136
Jerusalem from the top of the Dung Gate (George Lambert) *181*
Johnston's Gully 97, *143*
Johnston's Jolly 154, 198
Joyce, Staff Sergeant William 27, *27*, 43
  biographical note 225n2

Kantara 136
Karpura
  German train engine *176*
Karpura, Taurus Mountains (George Lambert) *168*
Kemal, Mustafa (Commander of Turkish Forces) 178
Kempe, Sergeant Humphrey 1
  biographical note 223n4 (chap. 1)
Kilia Liman 148, *153*
Kilid Bahr 50, *55*, *56*, 59, *68*, 139, 154, 162, *190*
King George V 6
Krithia 154
  second battle 158–9, 162
  terrain *125*

Lambert, Honorary Captain George xi, 4, 25, 29, 38, 43, 45, 47, 50, 57–8, 70, 87, 100, 105, 107, 112, 125, 154, 156, 160, 162, 164, 165, 178
  biographical note 223n7
  Gallipoli paintings 202–15
  later life 221
  painting in Palestine 17
  paints Gallipoli flora 82–3
  paints the Nek 80
  record of Abbassia General Hospital 186–7
  sketching at Tiberias *32*
  travelling with the Light Horse 128
  working methods 208–9
The landing, 25 April 1915 4, *11*, *13*, 74–9
Latham, Lieutenant Commander J.G. 46

Lawrence, Colonel T.E. ('Lawrence of Arabia') 178
Leane's Trench 92, *92*, 112, *113*
Legge Valley 64, 74, 197
Lemnos 50, *51*
Lieutenant General Sir Harry Chauvel, (George Lambert) *34*
Light Horse Officer in 14th Australian General Hospital, Abbassia (George Lambert) *185*
Lone Pine *7*, 70, 92, 100, 103, *103*, *104*, *138*, 139, 142, 143, *143*, *146*, 198, 215
Lone Pine, looking towards the Nek, Walker's Ridge (George Lambert) *156*
Loutit, Lieutenant Noel 198
  biographical note 228n12 (chap. 8)
  at Gun Ridge 197

Mackinnon, Lieutenant 164
Magdhaba, Egypt 128
Maidos *153*
Major Zeki Bey, Commander of Turkish Regiment at Gallipoli (George Lambert) *71*
Marine Trench 97
Masséna, run aground off Cape Helles Beach *124*
Milo, sunk off North Beach 116, 148, 152
Monash Gully 108, 112
Moore-Jones, Horace 11
  biographical note 224n21
Mount Lebanon *172*
Mudros, Armistice 41
Mudros Cemetery *50*
Mule Valley xii
Mungovan, R. 46

Narrows, the 59, 65, 133, *145*, *153*
National Collection of the Australian War Memorial 14
The Nek 1, 70, 79, 80, *107*, 154, 156, 210, *211*
The Nek, Walker's Ridge (George Lambert) *78*, 80
Nevinson, C.R.W. (artist) 22
North Beach 74

Ocean Beach, Gallipoli *152*, 198
Official History of the AIF
  first volume 195
  staff *191*

Official war histories 190–3
　the role of photography
　　197–202
Ottoman Empire (map) 66
Outposts
　No. 1 *115*, 148, *149*, *151*
　No. 2 *115*, *126*, *134*, 148, *151*
　Old No. 5 *126*
'Over the top' composite photograph
　by Frank Hurley *20*
Owen's Gully *138*, 198

Paget colour plates 96, 117
painting and the 'Gallipoli Spirit'
　202–15
Palestine 38, 128
Peace Conference (Palace of
　Versailles, Paris) 47
　Australian delegation *46*
Pearce, Senator George
　(Australian Minister for
　Defence) 10, 14
Péronne, northern France 43
photography
　equipment 133, 136
　and historical analysis 117
　role in the Historical Mission
　　21–2
picture models (dioramas) 203
plan models 198
Plugge's Plateau *76*, 77, 79, 112,
　150, *208*, *209*
Pope's Hill 10, 108, 139
Pope's Reserve Gully (Hill) *137*
Portrait of C.E.W. Bean, left profile
　(George Lambert) *184*
POWs on the Taurus Railway 177,
　177
Pozières (France) 159

Queen Elizabeth (British
　battleship) 55
Queensland Point Cemetery *15*
Quinn's Post *8*, 10, 105, 108, *110*,
　112, 117, *123*, 139, 142, *144*

Ravine Gully *122*
Razorback *77*, *126*, 148, *208*
Recruitment poster *4*
Rest Camp Taranto (George
　Lambert) *48*
Retaliation Farm Dressing Station,
　composite photograph by
　Frank Hurley *23*

Rhododendron Hill 108, 112
Rhododendron Spur *109*
River Clyde, run aground off
　Cape Helles 124, 154
River Clyde at Cape Helles
　(George Lambert) *163*
Robin, Corporal Philip 198
Robinson, Corporal
　(draughtsman) 191
Rogers, Sergeant George 39,
　45, 70, 79, 93, 104, 164,
　165, *191*
　biographical note 225n25
　later life 220
Romani, Egypt 38, 128
*Romani, Mount Royston in
　background* (George
　Lambert) *37*
Royal Australian Navy (RAN) 55
Russell's Top *77*, 79, 108, *137*,
　148, *149*, *151*

Salonica 136
Salt Lake 118, 148
Samothrace *91*
Sanders, Liman von (German
　commander) 59
Sari Bair Range 79, 112
Schuler, Phillip 5, *5*
2nd Australian Infantry Brigade
　162
Sedgwick, Lieutenant Albert
　(24th Battalion) 1918 *28*
7th Australian Light Horse 66,
　68–9, 136, 139
7th Light Horse 68
Shell Green *140*, *141*, 142
Shrapnel Gully 70, *99*, 112, 139,
　198
Silas, Ellis 1
　biographical note 224n21
Silt Spur *119*, 119–20, *120*
Sinai 38, 128
Smart, Captain Henry Casimir
　24, 43
　biographical note 225n3
Sniper's Nest 112, 116, *116*, 118,
　118
South African War ix
The Sphinx *77*, 79, *150*, *208*
The Sphinx from Plugge's Plateau
　(George Lambert) *81*
Spooks Plateau 112
Spruce, Lance Corporal William
　(7th Light Horse) 70, *131*,
　139, 154

biographical note 226n12
Steele's Post *18*
Storm effect, Malta, (George
　Lambert) *49*
Strazeele, Flanders (now France) 26
Study for 'ANZAC, the landing, 1915'
　(George Lambert) *206*, *207*
*Study for dead trooper and detail
　of Turkish trench, Gallipoli
　(Pro patria)* (George
　Lambert) *157*
*Study for 'The charge of the 3rd
　Light Horse Brigade at the
　Nek, 7 August 1915'*
　(George Lambert) *210*
Study of Arbutus shrub
　(George Lambert) *83*
Sunset, Lemnos (George Lambert)
　51
Surprise Gully *121*
Suvla Bay 112, 118, 148
Swanston, Sergeant William xi, 4,
　65, 70, 100, 127, 128, 138, 139,
　142, 148, 154, 182, 215
　background in photography 136
　biographical note 223n8
　later life 221–2
　partnership with William James
　　136
　working method 139

Table Top plateau *126*, 148
Taranto, Southern Italy 47
Tasmania Post 92
Taurus Mountains *167*
Taurus Railway 165–78, *166*
　and Australian POWs 177
10th (Western Australian)
　Regiment at the Nek 79
*The charge of the 3rd Light Horse
　Brigade at the Nek, 7 August
　1915* (George Lambert)
　211–5, *212–13*
*The Nek, Walker's Ridge, site of the
　charge of the Light Horse*
　(George Lambert) *78*
*The silver lining – sunset over
　Imbros as seen from ANZAC*
　(Charles Bean) *16*
*The Sphinx, from Suvla side, grey
　day* (George Lambert) *85*
*The Sphinx from Plugge's Plateau,
　Gallipoli* (George Lambert)
　81
The top of the Taurus Mountains
　(George Lambert) *168*

3rd Australian Brigade at 400
　Plateau 198
3rd Australian Light Horse
　Brigade at the Nek 38, 79
39th Battalion, Houplines, France
　1916 *39*
Tommies' Trench 102, *102*, 158,
　158, 162
Treloar, Captain John 14, 17, 128,
　133, 182, 215
　biographical note 224n24
Turkish and Arab soldiers
　(demobilised) *170*, *171*, *174*,
　175, *179*
Turkish monument at the Nek *211*
Turkish trenches *7*
28th British Infantry Division 57,
　65, 69
24th Australian Battalion *28*, 31

Valletta, Malta 47
Valley of Despair *121*

*Wadi bed between El Arish and
　Magdhaba* (George Lambert)
　35
Walker's Ridge xii, 148, *149*, *150*
Wanliss Gully *121*
Wellington, H. 112
West Mudros rest camp, Lemnos
　(George Lambert) *60*
Western Front 9, 11, 17, 41
Whyte, Archie 5
Wilkins, Hubert viii, xi, 4, 14, 25,
　26, *26*, 27, *27*, 29, 38, 45, 50,
　54, 57, 70, 74, 79, *95*, 103,
　105, 107, 108, 116–7, 127,
　154, 162, 164, 165, 182, 197,
　215
　biographical note 224n29
　later life 222
　notes on Essad Pasha's
　　Headquarters 197
　at the Western Front 100
　working method 95–100, 139
Wire Gully *97*
Wood, Miss 46
Woolley, Sapper Arthur 66, 74, *94*
　biographical note 226n9

Zeki, Major (commander of 1st
　Battalion of the 57th Turkish
　Regiment) 70, 100, 104,
　106, 112